Physics
Reference Table
Workbook

Authors:
Ron Pasto – Retired Physics Teacher
William Docekal – Retired Science Teacher

About This Workbook –

Many questions on the New York State Physical Setting/PHYSICS Regents exam may be answered simply by using information given on the Reference Table. Other questions may require information from the Reference Table to set-up calculations in order to determine the answer. Knowing what information is on the Reference Table and where to find it is a very important step towards being successful on the Regents exam.

© 2011, Topical Review Book Company, Inc. All rights reserved.
P. O. Box 328
Onsted, MI 49265-0328
www.topicalrbc.com

Physics Reference Table Workbook

The Introduction – Overview, The Chart and Additional Information –

This workbook contains 36 sections, 26 dealing with the equations and 10 dealing with charts. In each section, carefully read the introduction material. Read and understand the example given for that equation and its solution. Study the *Additional Information* given. When you have a solid knowledge of that section's material, move on to the topic questions.

Set 1 – Questions and Answers –

Set 1 questions will test your understanding of that particular section. Do all questions in Set 1 and then correct your work by going to the answers for Set 1, which are at the end of the section. The explanation given will help you to understand any mistakes you may have made. If not, ask you teacher for additional help.

Set 2 – Questions –

The answers to these questions are in a separate answer key for you to use in checking your answers. Correctly answering these questions will show you and your teacher that you understand the subject matter for that particular section.

All of us at Topical Review Book Company hope that a complete understanding of the Physics Reference Table will help to increase your knowledge of Physics and that your grade will improve.

Good luck in Physics and on the Regents or the school exam. – The authors.

Physical Setting/Physics
Reference Table Booklet
Table of Contents

Mechanics – Kinematics

Mechanics may be defined as the branch of physics that studies the motion, the causes of the motion and changes in the motion of objects. It is usually divided into two parts: kinematics and dynamics.

Kinematics: Kinematics is the study of the types of motion an object may have. Five equations are used in this area.

Speed is the rate of change of position or rate of motion of an object. Velocity is the rate of motion of an object in a particular direction. Speed is the magnitude of the velocity vector.

Average speed is calculated by dividing the total distance traveled by the time needed to travel that distance. Average velocity is displacement divided by time.

$$\bar{v} = \frac{d}{t}$$

where: $\bar{v}$ = average velocity or average speed
d = displacement or distance
t = time interval

Example: A person observes a fireworks display from a safe distance of 0.750 kilometer. Assuming that sound travels at 340. meters per second in air, what is the time between the person seeing and hearing a fireworks explosion?

(1) 0.453 s
(2) 2.21 s
(3) 410. s
(4) 2.55×10^5 s

Solution: 2   Convert 0.750 km to meters to be consistent with the units given for the speed of sound. The equation gives: 340. m/s = (750. m)/t. Solving, t = 2.21 s.

Acceleration is the rate of change of velocity of an object. It is calculated by dividing the change in velocity of the object by the time needed to make that change.

$$a = \frac{\Delta v}{t}$$

where: a = acceleration
Δv = change in velocity or speed
t = time interval

Example: An observer recorded the following data for the motion of a car undergoing constant acceleration. What was the magnitude of the acceleration of the car?

Time (s)	Speed (m/s)
3.0	4.0
5.0	7.0
6.0	8.5

(1) 1.3 m/s²
(2) 2.0 m/s²
(3) 1.5 m/s²
(4) 4.5 m/s²

Solution: 3   Using any time interval and the corresponding change in velocity gives an acceleration of 1.5 m/s².

For example: $a = \dfrac{(8.5 \text{ m/s} - 4.0 \text{ m/s})}{(6.0 \text{ s} - 3.0 \text{ s})} = \dfrac{4.5 \text{ m/s}}{3.0 \text{ s}} = 1.5 \text{ m/s}^2$

If the initial speed or velocity (v_i) is known, the final speed or velocity (v_f) may be calculated if the acceleration and time are known. If the object starts from rest, $v_i = 0$.

$$v_f = v_i + at$$

where: v_i = initial velocity or speed
v_f = final velocity or speed
a = acceleration
t = time interval

Example: An object is dropped from rest and falls freely 20. meters to Earth. When is the speed of the object 9.8 meters per second?

(1) during the entire first second of its fall

(2) at the end of its first second of fall

(3) during its entire time of fall

(4) after it has fallen 9.8 meters

Solution: 2   The acceleration of a freely falling object is the acceleration due to gravity (g), found on the List of Physical Constants in the reference table. The object starts from rest and its speed increases during each second it falls. Substitution in the equation using $t = 1.0$ s gives $v_f = 0 + (9.81 \text{ m/s}^2)(1.0 \text{ s})$. Solving, $v_f = 9.81$ m/s.

If the initial speed or velocity, acceleration and time of travel are known, the distance traveled in that time can be calculated. Again, if the object starts from rest, $v_i = 0$.

$$d = v_i t + \frac{1}{2}at^2$$

where: d = displacement or distance
v_i = initial velocity or speed
a = acceleration
t = time interval

Example: A ball is thrown horizontally at a speed of 24 meters per second from the top of a cliff. If the ball hits the ground 4.0 seconds later, approximately how high is the cliff?

(1) 6.0 m   (2) 39 m   (3) 78 m   (4) 96 m

Solution: 3   The horizontal and vertical motion of the ball are independent of each other. The ball falls vertically for 4.0 s before it strikes the ground and has an acceleration equal to that due to gravity, found on the List of Physical Constants. Since the ball starts from rest in the vertical, $v_i = 0$. Substituting into the equation using $a = g$, $d = 0 + \frac{1}{2}(9.81 \text{ m/s}^2)(4.0 \text{ s})^2 = 78$ m.

If the initial speed or velocity, acceleration and distance traveled are known, the final speed or velocity may be calculated. Again, if the object starts from rest, $v_i = 0$.

$$v_f^2 = v_i^2 + 2ad$$

where: v_i = initial velocity or speed
v_f = final velocity or speed
a = acceleration
d = displacement or distance

Example: An object with an initial speed of 4.0 meters per second accelerates uniformly at 2.0 meters per second^2 in the direction of its motion for a distance of 5.0 meters. What is the final speed of the object?

(1) 6.0 m/s   (2) 10. m/s   (3) 14 m/s   (4) 36 m/s

Solution: 1   Substituting into the equation: $v_f^2 = (4.0 \text{ m/s})^2 + 2(2.0 \text{ m/s}^2)(5.0 \text{ m})$. Solving for v_f gives 6.0 m/s.

Kinematics - Additional Information:

- When the speed is constant, it is the average speed.

- If the speed of an object changes, average speed may be calculated as follows: $\bar{v} = \dfrac{(v_i + v_f)}{2}$.
- The change in velocity (Δv) is calculated $\Delta v = v_f - v_i$.

- In the case of free fall and projectile motion, the vertical acceleration is that due to gravity (g). In the equations, g may then be substituted for a. The value of g is on the List of Physical Constants.

- In some questions, it is important to distinguish between scalar quantities (those having magnitude or size only) and vector quantities (those having both magnitude and direction). Speed and distance are scalars. Velocity, displacement and acceleration are vectors.

- When an object is increasing its speed, the acceleration is a positive quantity. When slowing down, the acceleration is negative (deceleration).

1. At an outdoor physics demonstration, a delay of 0.50 second was observed between the time sound waves left a loudspeaker and the time these sound waves reached a student through the air. If the air is at STP, how far was the student from the speaker?

   (1) 1.5×10^{-3} m
   (2) 1.7×10^2 m
   (3) 6.6×10^2 m
   (4) 1.5×10^8 m   1 _____

2. Approximately how much time does it take light to travel from the Sun to Earth?

   (1) 2.00×10^{-3} s
   (2) 1.28×10^0 s
   (3) 5.00×10^2 s
   (4) 4.50×10^{19} s   2 _____

3. The speed of a car is increased uniformly from 20. meters per second to 30. meters per second in 4.0 seconds. The magnitude of the car's average acceleration in this 4.0-second interval is

   (1) 0.40 m/s²   (3) 10 m/s²
   (2) 2.5 m/s²   (4) 13 m/s²   3 _____

4. Velocity is to speed as displacement is to

   (1) acceleration   (3) momentum
   (2) time   (4) distance   4 _____

5. An astronaut drops a hammer from 2.0 meters above the surface of the Moon. If the acceleration due to gravity on the Moon is 1.62 meters per second², how long will it take for the hammer to fall to the Moon's surface?

   (1) 0.62 s   (3) 1.6 s
   (2) 1.2 s   (4) 2.5 s   5 _____

6. A rocket initially at rest on the ground lifts off vertically with a constant acceleration of 2.0×10^1 meters per second². How long will it take the rocket to reach an altitude of 9.0×10^3 meters?

   (1) 3.0×10^1 s   (3) 4.5×10^2 s
   (2) 4.3×10^1 s   (4) 9.0×10^2 s   6 _____

7. A car initially traveling at a speed of 16 meters per second accelerates uniformly to a speed of 20. meters per second over a distance of 36 meters. What is the magnitude of the car's acceleration?

   (1) 0.11 m/s²   (3) 0.22 m/s²
   (2) 2.0 m/s²   (4) 9.0 m/s²   7 _____

8. A stream is 30. m wide and its current is flowing at 1.5 m/s. A boat is launched with a velocity of 2.0 m/s eastward from the west bank of the stream. How much time is required for the boat to reach the opposite bank of the stream?

   (1) 8.6 s   (3) 15 s
   (2) 12 s   (4) 60. s   8 _____

9. The speed of an object undergoing constant acceleration increases from 8.0 meters per second to 16.0 meters per second in 10. seconds. How far does the object travel during the 10. seconds?

(1) 3.6×10^2 m   (3) 1.2×10^2 m
(2) 1.6×10^2 m   (4) 8.0×10^1 m   9 _____

10. A rock falls from rest a vertical distance of 0.72 meter to the surface of a planet in 0.63 second. The magnitude of the acceleration due to gravity on the planet is

(1) 1.1 m/s²   (3) 3.6 m/s²
(2) 2.3 m/s²   (4) 9.8 m/s²   10 _____

11. A projectile has an initial horizontal velocity of 15 meters per second and an initial vertical velocity of 25 meters per second. Determine the projectile's horizontal displacement if the total time of flight is 5.0 seconds. [Neglect friction.] [Show all work, including the equation and substitution with units.]

_____ m

Base your answers to question 12 a and b on the information below.

A car traveling at a speed of 13 meters per second accelerates uniformly to a speed of 25 meters per second in 5.0 seconds.

12. a) Calculate the magnitude of the acceleration of the car during this 5.0-second time interval. [Show all work, including the equation and substitution with units.]

b) A truck traveling at a constant speed covers the same total distance as the car in the same 5.0-second time interval. Determine the speed of the truck. [Show all work, including the equation and substitution with units.]

b)_____ m/s

Base your answers to questions 13 *a*, *b*, and *c* on the information below.

   A car on a straight road starts from rest and accelerates at 1.0 meter per second² for
10. seconds. Then the car continues to travel at constant speed for an additional 20. seconds.

13. *a*) Determine the speed of the car at the end of the first 10. seconds. _____ m/s

   b) On the accompanying grid, use a ruler or
   straightedge to construct a graph of the
   car's speed as a function of time for the
   entire 30.-second interval.

   c) Calculate the distance the car travels in
   the first 10.-seconds. [Show all work,
   including the equation and substitution
   with units.]

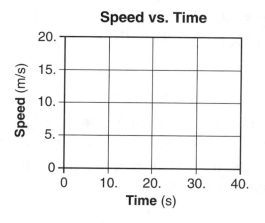

Base your answers to questions 14 *a* and *b* on the information below.

   A physics class is to design an experiment to determine the acceleration of a student on inline
skates coasting straight down a gentle incline. The incline has a constant slope. The students have
tape measures, traffic cones, and stopwatches.

14. *a*) Describe a procedure to obtain the measurements necessary for this experiment.

   _____

   _____

   _____

   _____

   b) Indicate which equation(s) they should use to determine the student's acceleration.

15. A 2.00×10^6-hertz radio signal is sent a distance of 7.30×10^{10} meters from Earth to a spaceship orbiting Mars. Approximately how much time does it take the radio signal to travel from Earth to the spaceship?

(1) 4.11×10^{-3} s
(2) 2.19×10^8 s
(3) 2.43×10^2 s
(4) 1.46×10^{17} s   15 _____

16. A projectile is fired from a gun near the surface of Earth. The initial velocity of the projectile has a vertical component of 98 meters per second and a horizontal component of 49 meters per second. How long will it take the projectile to reach the highest point in its path?

(1) 5.0 s   (3) 20. s
(2) 10. s   (4) 100. s   16 _____

17. A roller coaster, traveling with an initial speed of 15 meters per second, decelerates uniformly at –7.0 meters per second^2 to a full stop. Approximately how far does the roller coaster travel during its deceleration?

(1) 1.0 m   (3) 16 m
(2) 2.0 m   (4) 32 m   17 _____

18. An astronaut standing on a platform on the Moon drops a hammer. If the hammer falls 6.0 meters vertically in 2.7 seconds, what is its acceleration?

(1) 1.6 m/s^2 (3) 4.4 m/s^2
(2) 2.2 m/s^2 (4) 9.8 m/s^2 18 _____

19. On a highway, a car is driven 80. kilometers during the first 1.00 hour of travel, 50. kilometers during the next 0.50 hour, and 40. kilometers in the final 0.50 hour. What is the car's average speed for the entire trip?

(1) 45 km/h   (3) 85 km/h
(2) 60. km/h   (4) 170 km/h   19 _____

Base your answers to question 20 on the information below.

The combined mass of a race car and its driver is 600. kilograms. Traveling at constant speed, the car completes one lap around a circular track of radius 160 meters in 36 seconds.

20. Calculate the speed of the car. [Show all work, including the equation and substitution with units.]

21. During a 5.0-second interval, an object's velocity changes from 25 meters per second east to 15 meters per second east. Determine the magnitude and direction of the object's acceleration.

_____ m/s² _____

Base your answers to questions 22 *a*, *b* and *c* on the information and diagram below.

A projectile is launched horizontally at a speed of 30. meters per second from a platform located a vertical distance *h* above the ground. The projectile strikes the ground after time *t* at horizontal distance *d* from the base of the platform. [Neglect friction.]

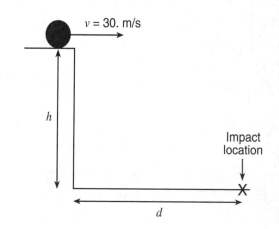

22. *a*) On the accompanying diagram, sketch the theoretical path of the projectile.

b) Calculate the horizontal distance, *d*, if the projectile's total time of flight is 2.5 seconds. [Show all work, including the equation and substitution with units.]

c) Express the projectile's total time of flight, *t*, in terms of the vertical distance, *h*, and the acceleration due to gravity, *g*. [Write an appropriate equation and solve it for *t*.]

1. 2 Locate the equation $\bar{v} = d/t$. The speed of sound in air at STP is found in the reference table on the List of Physical Constants. Since it is constant, it is the average speed. Substituting into the equation gives 3.31×10^2 m/s $= d/(0.5$ s$)$. Solving, $d = 1.7 \times 10^2$ m.

2. 3 Under Mechanics, find the equation $\bar{v} = d/t$. The speed of light is found on the List of Physical Constants. Since it is a constant, it is the average speed. The mean distance from the Earth to the Sun is also found on the List of Physical Constants. Solving for t and substituting gives
$$t = \frac{(1.50 \times 10^{11} \text{ m})}{(3.00 \times 10^8 \text{ m/s})} = 5.00 \times 10^2 \text{ s}.$$

3. 2 The change in speed (Δv) can be expressed as $v_f - v_i$. This gives $\Delta v = 30.$ m/s $- 20.$ m/s $= 10.$ m/s. Substitution into the equation $a = \Delta v/t$ for acceleration gives $a = (10.$ m/s$)/(4.0$ s$)$. Solving gives $a = 2.5$ m/s^2. Assuming the acceleration is constant, this is also the average acceleration.

4. 4 A vector is a quantity possessing both magnitude (size) and direction. Velocity is a vector defined as a speed measured in a particular direction. Speed is the magnitude of the velocity vector. Displacement is a vector defined as a distance measured in a particular direction. Distance is the magnitude of the displacement vector. By analogy, the correct answer is distance.

5. 3 The equation to use is $d = v_i t + \frac{1}{2}at^2$. The value of a will be the acceleration due to gravity at the surface of the Moon. Assume the hammer is dropped from rest. Substituting into the equation gives 2.0 m $= 0 + \frac{1}{2} (1.62$ m/s$^2)(t^2)$. Solving, $t = 1.6$ s.

6. 1 Under Mechanics find the equation $d = v_i t + \frac{1}{2} at^2$. Substitution into the equation with $v_i = 0$ (the object starts from rest) gives: 9.0×10^3 m $= \frac{1}{2} (2.0 \times 10^1$ m/s$^2)(t^2)$. Solving, $t = 3.0 \times 10^1$ s.

7. 2 Find the equation $v_f^2 = v_i^2 + 2ad$. Substitution gives $(20.$ m/s$)^2 = (16$ m/s$)^2 + 2a(36$ m$)$. Solving for a yields 2.0 m/s^2.

8. 3 Under Mechanics in the reference table, find the equation $\bar{v} = d/t$. $\bar{v}$ is the constant eastern velocity of the boat and d is the width of the river. Substituting gives 2.0 m/s $= (30.$ m$)/t$. Solving, $t = 15$ s.

9. 3   Two equations are needed: $a = \frac{\Delta v}{t}$ and $v_f^2 = v_i^2 + 2ad$.

Using the acceleration equation, $a = \frac{(16.0 \text{ m/s} - 8.0 \text{ m/s})}{(10. \text{ s})} = 0.80$ m/s^2.

Substituting into the second equation gives
$(16.0 \text{ m/s})^2 = (8.0 \text{ m/s})^2 + 2(0.80 \text{ m/s}^2)(d)$. Solving, $d = 1.2 \times 10^2$ m

or

Under Mechanics, find the equation $\bar{v} = d/t$. To find the average

speed, $\bar{v} = \frac{(v_i + v_f)}{2}$. Substituting, $\bar{v} = \frac{(8.0 \text{ m/s} + 16.0 \text{ m/s})}{2} = 12.0$ m/s.

Substitution into the first equation gives $(12.0 \text{ m/s}) = \frac{d}{(10. \text{ s})}$.

Solving, $d = 1.2 \times 10^2$ m

10. 3   Find the equation $d = v_i t + \frac{1}{2}at^2$.
Since the rock starts from rest, $v_i = 0$.

Substitution gives $0.72 \text{ m} = 0 + \frac{1}{2}(a)(0.63 \text{ s})^2$.   Solving, $a = 3.6$ m/s^2.

11.   Answer: 75 m

Explanation: The horizontal displacement of the projectile depends only upon the horizontal velocity and the total time of flight. Neglecting friction, the magnitude of the horizontal velocity remains constant during the flight. Under Mechanics, find the equation $\bar{v} = \frac{d}{t}$.
Substitution into the equation gives $(15 \text{ m/s}) = d/(5.0 \text{ s})$. Solving for d gives 75 m.

12.   a) $a = \frac{\Delta v}{t}$ Explanation: Given the initial speed, final speed and time, the

$a = \frac{25 \text{ m/s} - 13 \text{ m/s}}{5.0 \text{s}}$ acceleration may be calculated.

$a = 2.4 \text{ m/s}^2$

b) Example of acceptable response:

$d = v_i t + \frac{1}{2}at^2$ $d = (13 \text{ m/s})(5.0 \text{ s}) + \frac{1}{2}(2.4 \text{ m/s}^2)(5.0 \text{ s})^2 = 95$ m

$\bar{v} = \frac{d}{t}$ $\bar{v} = \frac{95 \text{ m}}{5.0 \text{ s}} = 19$ m/s

Explanation: First, calculate the distance the car travels during the 5.0 seconds using the acceleration calculated in 12*a* (2.4 m/s^2). This is the distance the truck travels during this time. Knowing the distance and time, the speed of the truck may be calculated. Since the truck is traveling at a constant speed, the average speed and the truck's constant speed are the same.

13.　a) Answer: 10 m/s

　　　Explanation: Under Mechanics, find the equation $v_f = v_i + at$. $v_i = 0$ since the car starts from rest. Substituting into the equation gives $v_f = (1.0 \text{ m/s}^2)(10. \text{ s}) = 10. \text{ m/s}$.

b)

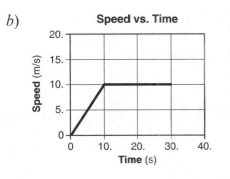

Speed vs. Time

Explanation: From 0 s to 10. s, the cars speed increases uniformly from 0 m/s to 10. m/s. From 10. s to 30. s, the car travels at a constant speed of 10. m/s.

c)　$d = v_i t + \dfrac{1}{2} at^2$ 　　　　　　　　$d = \text{area} = \dfrac{1}{2} bh$

　　$d = 0 + \dfrac{1}{2}(1.0 \text{ m/s}^2)(10. \text{ s})^2$ 　*or*　 $d = \dfrac{1}{2}(10. \text{ s})(10. \text{ m/s})$

　　$d = 50. \text{ m}$ 　　　　　　　　　　　　　$d = 50. \text{ m}$

Explanation: Use the equation $d = v_i t + \frac{1}{2} at^2$.
Substitute into the equation using $v_i = 0$ and calculate d.
or
The distance traveled during the first 10. s is equal to the area under the triangle on the graph from 0 s to 10. s. Under Geometry and Trigonometry in the reference table, find the equation for the area of a triangle ($A = \frac{1}{2} bh$). The base of the triangle is 10. s and the height is 10. m/s. Calculate A, which is the distance traveled.

14.　　a) The class must measure the time needed for the student to travel a given measured distance.

　　　b) $d = v_i t + \frac{1}{2} at^2$

　　　　　　　or

　　　　$a = \dfrac{2d}{t^2}$

Explanation: If the student starts from rest, $v_i = 0$. With the measured values of d and t, the acceleration of the student can be calculated.

Mechanics　　　　　　　　　　　　　　**Page 11**

Mechanics – Dynamics

Dynamics: Dynamics deals with forces and how they produce or affect motion. This brings into play friction, gravity, momentum, circular motion, work and energy.

Newton's Second Law: This law relates the net force acting on an object and the change in motion (acceleration) it produces.

The acceleration of an object is calculated by dividing the net or unbalanced force acting on the object by the object mass.

$$a = F_{net}/m$$

where: a = acceleration
F_{net} = net force
m = mass

Example: A 25-newton horizontal force northward and a 35-newton horizontal force southward act concurrently on a 15-kilogram object on a frictionless surface. What is the magnitude of the object's acceleration?

(1) 0.67 m/s²   (2) 1.7 m/s²   (3) 2.3 m/s²   (4) 4.0 m/s²

Solution: 1 The net force acting on the object is the difference between the oppositely directed 25 N and 35 N forces, or 10. N. Substitution into the equation gives $a = (10. N)/(15 kg)$. Solving, $a = 0.67$ m/s².

Dynamics - Additional Information:

• Newton's First Law states that an object at rest or in uniform motion tends to stay at rest or in uniform motion unless acted upon by an unbalanced or net force. Newton's Third Law states that for every action, there is an equal but opposite reaction.

• Inertia is the property of matter that resists a change in its state of rest or uniform motion. Mass is a quantitative measure of the inertia of an object. The greater the mass, the greater the inertia of an object.

• In the above equation, the force must be the net force (vector sum of all forces) acting on the object.

• If the net force acting on the object is in the same direction as the motion of the object, the speed of the object increases. It produces a positive acceleration.

• If the net force acting on the object is in the opposite direction to the motion of the object, the speed of the object decreases. It produces a negative acceleration or a deceleration.

• If the net force acts at a right angle to the motion, it changes only the direction of the motion (see Circular Motion).

Set 1 — Dynamics

1. A net force of 25 newtons is applied horizontally to a 10.-kilogram block resting on a table. What is the magnitude of the acceleration of the block?

   (1) 0.0 m/s²   (3) 0.40 m/s²
   (2) 0.26 m/s²   (4) 2.5 m/s²   1 _____

2. A force of 1 newton is equivalent to 1

   (1) $\dfrac{\text{kg} \bullet \text{m}}{\text{s}^2}$ (3) $\dfrac{\text{kg} \bullet \text{m}^2}{\text{s}^2}$

   (2) $\dfrac{\text{kg} \bullet \text{m}}{\text{s}}$ (4) $\dfrac{\text{kg}^2 \bullet \text{m}^2}{\text{s}^2}$ 2 _____

3. A net force of 10. newtons accelerates an object at 5.0 meters per second². What net force would be required to accelerate the same object at 1.0 meter per second² ?

   (1) 1.0 N   (3) 5.0 N
   (2) 2.0 N   (4) 50. N   3 _____

4. A 1.5-kilogram lab cart is accelerated uniformly from rest to a speed of 2.0 meters per second in 0.50 second. What is the magnitude of the force producing this acceleration?

   (1) 0.70 N   (3) 3.0 N
   (2) 1.5 N   (4) 6.0 N   4 _____

5. The accompanying diagram shows a force of magnitude F applied to a mass at angle θ relative to a horizontal frictionless surface.

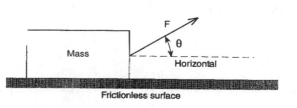

   As angle θ is increased, the horizontal acceleration of the mass
   (1) decreases   (2) increases   (3) remains the same   5 _____

6. A horizontal force of 8.0 newtons is used to pull a 20. kilogram wooden box on a frictionless surface.

   Calculate the magnitude of the acceleration of the box. [Show all work, including the equation and substitution with units.]

7. A constant unbalanced force is applied to an object for a period of time. Which graph best represents the acceleration of the object as a function of elapsed time?

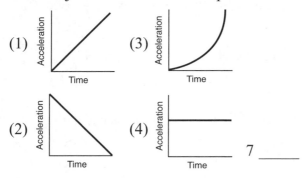

(1)　　　(3)

(2)　　　(4)

7 _____

8. As a car is driven south in a straight line with decreasing speed, the acceleration of the car must be

(1) directed northward
(2) directed southward
(3) zero
(4) constant, but not zero

8 _____

9. The diagram below shows a horizontal 8.0-newton force applied to a 4.0-kilogram block on a frictionless table. What is the magnitude of the block's acceleration?

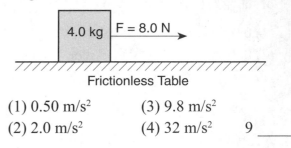

Frictionless Table

(1) 0.50 m/s²　　　(3) 9.8 m/s²
(2) 2.0 m/s²　　　(4) 32 m/s²

9 _____

10. A 2.0-kilogram body is initially traveling at a velocity of 40. meters per second east. If a constant force of 10. newtons due east is applied to the body for 5.0 seconds, the final speed of the body is

(1) 15 m/s　　　(3) 65 m/s
(2) 25 m/s　　　(4) 130 m/s

10 _____

Base your answers to questions 11 *a* and *b* on the information and diagram below.

A 10.-kilogram box, sliding to the right across a rough horizontal floor, accelerates at –2.0 meters per second² due to the force of friction.

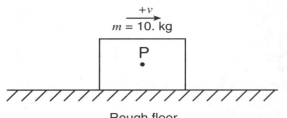

Rough floor

11. *a*) Calculate the magnitude of the net force acting on the box. [Show all work, including the equation and substitution with units.]

b) On the diagram above, draw a vector representing the net force acting on the box. Begin the vector at point *P* and use a scale of 1.0 centimeter = 5.0 newtons.

1. 4   This is an application of Newton's Second Law. Under Mechanics, find the equation $a = \dfrac{F_{net}}{m}$.

      Substitution into the equation gives $a = \dfrac{(25 \text{ N})}{(10. \text{ kg})}$. Solving gives $a = 2.5 \text{ m/s}^2$.

2. 1   The unit of force is based on the units of Newton's Second Law which is $a = F_{net}/m$. Solving for F_{net}

      gives $F_{net} = ma$. The unit of force is the product of the units of mass (kg) and acceleration (m/s^2).

      This is kg $\bullet \dfrac{\text{m}}{\text{s}^2}$, which is the Newton (N).

3. 2   Use $a = \dfrac{F_{net}}{m}$ to find the mass of the first object. Substituting into the equation gives

      $5.0 \text{ m/s}^2 = (10. \text{ N})/m$ and solving for m gives 2.0 kg. Now use the equation to solve for the

      acceleration of the second object using a mass of 2.0 kg. Substituting into the equation gives

      $1.0 \text{ m/s}^2 = \dfrac{F_{net}}{(2.0 \text{ kg})}$ and solving gives $F_{net} = 2.0 \text{ N}$.

4. 4   Two equations are needed: $a = \dfrac{\Delta v}{t}$ and $a = \dfrac{F_{net}}{m}$. Solving for a gives

      $a = \dfrac{(2.0 \text{ m/s})}{(0.50 \text{ s})} = 4.0 \text{ m/s}^2$. Use this acceleration in the second equation.

      Then $4.0 \text{ m/s}^2 = \dfrac{F_{net}}{(1.5 \text{ kg})}$ and $F_{net} = 6.0 \text{ N}$.

5. 1   As the angle θ increases, the horizontal component of force F decreases. It is this component that produces the horizontal acceleration of the mass. Under Mechanics, find the equation $a = F_{net}/m$. This shows that the acceleration varies directly with the force producing the acceleration. As this force decreases, the acceleration decreases.

6.   Answer: 0.40 m/s^2

    Explanation: The equation needed is $a = F_{net}/m$. Substitution in gives $a = 8.0 \text{ N}/20 \text{ kg}$. Solving, $a = 0.40 \text{ m/s}^2$.

Gravity: Gravity is the is the mutual force of attraction between all particles of matter. This force varies directly with the masses of the objects and inversely with the square of the distance between their centers. *G* in the equation is the constant of proportionality and is called the universal gravitational constant.

$$F_g = \frac{Gm_1m_2}{r^2}$$

where: F_g = weight or force due to gravity
G = universal gravitational constant
m = mass
r = radius or distance between centers

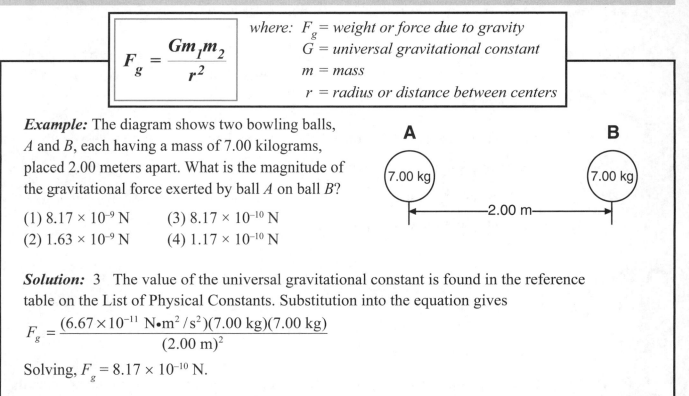

Example: The diagram shows two bowling balls, *A* and *B*, each having a mass of 7.00 kilograms, placed 2.00 meters apart. What is the magnitude of the gravitational force exerted by ball *A* on ball *B*?

A   7.00 kg   B   7.00 kg
|———2.00 m———|

(1) 8.17×10^{-9} N   (3) 8.17×10^{-10} N
(2) 1.63×10^{-9} N   (4) 1.17×10^{-10} N

Solution: 3   The value of the universal gravitational constant is found in the reference table on the List of Physical Constants. Substitution into the equation gives

$$F_g = \frac{(6.67 \times 10^{-11} \text{ N} \cdot \text{m}^2 / \text{s}^2)(7.00 \text{ kg})(7.00 \text{ kg})}{(2.00 \text{ m})^2}$$

Solving, $F_g = 8.17 \times 10^{-10}$ N.

The force of gravity acting on an object causes it to accelerate when dropped in a gravitational field. From Newton's Second Law, this acceleration (g) is equal to the gravitational force acting on the object (or the weight of the object) divided by the mass of the object.

$$g = \frac{F_g}{m}$$

where: g = accelaration due to gravity or gravitational field strength
F_g = weight or force due to gravity
m = mass

Example: A 2.0-kilogram object is falling freely near Earth's surface. What is the magnitude of the gravitational force that Earth exerts on the object?

(1) 20. N   (2) 2.0 N   (3) 0.20 N   (4) 0.0 N

Solution: 1   The value of *g* is given on the List of Physical Constants in the reference table. Substitution into the equation gives 9.81 m/s² = F_g/(2.0 kg).

Solving, F_g = 20. N.

Gravity - Additional Information:

- When the masses are the Earth, Moon or Sun, their masses are found on the List of Physical Constants.

- F_g in either equation is the weight of the object.

- The value of G (universal gravitational constant) and g (acceleration due to gravity) are found on the List of Physical Constants.

Set 1 — Gravity

1. The acceleration due to gravity on the surface of planet X is 19.6 meters per second². If an object on the surface of this planet weighs 980. newtons, the mass of the object is

   (1) 50.0 kg   (3) 490. N
   (2) 100. kg   (4) 908 N   1 _____

2. The centers of two 15.0-kilogram spheres are separated by 3.00 meters. The magnitude of the gravitational force between the two spheres is approximately

   (1) 1.11×10^{-10} N
   (2) 3.34×10^{-10} N
   (3) 1.67×10^{-9} N
   (4) 5.00×10^{-9} N   2 _____

3. As a meteor moves from a distance of 16 Earth radii to a distance of 2 Earth radii from the center of Earth, the magnitude of the gravitational force between the meteor and Earth becomes

   (1) $\frac{1}{8}$ as great
   (2) 8 times as great
   (3) 64 times as great
   (4) 4 times as great   3 _____

4. What is the acceleration due to gravity at a location where a 15.0-kilogram mass weighs 45.0 newtons?

   (1) 675 m/s²   (3) 3.00 m/s²
   (2) 9.81 m/s²   (4) 0.333 m/s²   4 _____

5. The diagram below shows a 5.00-kilogram block at rest on a horizontal, frictionless table.

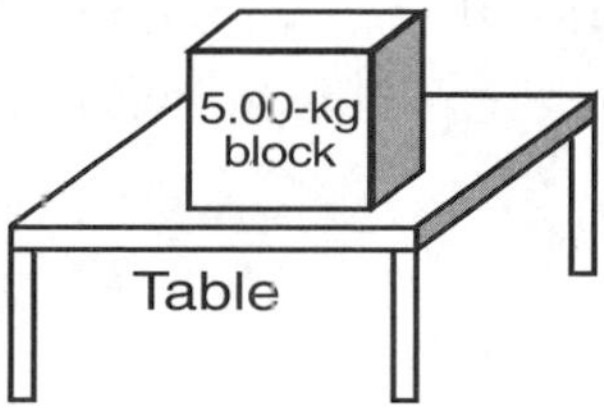

   Which diagram best represents the force exerted on the block by the table?

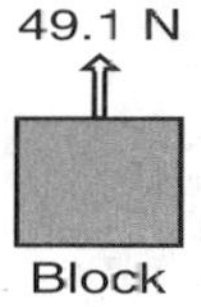

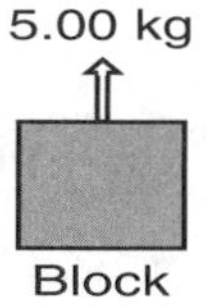

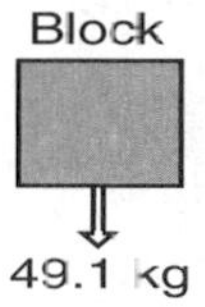

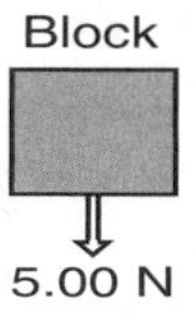

   5 _____

6. A 60-kilogram skydiver is falling at a constant speed near the surface of Earth. The magnitude of the force of air friction acting on the skydiver is approximately

(1) 0 N   (3) 60 N
(2) 6 N   (4) 600 N   6 _____

Note: Question 7 has only three choices.

7. If the magnitude of the gravitational force of Earth on the Moon is F, the magnitude of the gravitational force of the Moon on Earth is

(1) smaller than F
(2) larger than F
(3) equal to F 7 _____

8. Which diagram best represents the gravitational forces, F_g, between a satellite, S, and Earth?

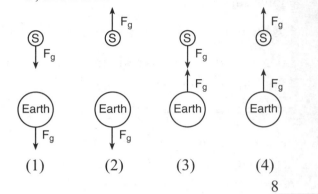

(1)   (2)   (3)   (4)
                                          8 _____

9. Earth's mass is approximately 81 times the mass of the Moon. If Earth exerts a gravitational force of magnitude F on the Moon, the magnitude of the gravitational force of the Moon on Earth is

(1) F (2) $\frac{F}{81}$ (3) $9F$ (4) $81F$ 9 _____

---

10. The gravitational force of attraction between Earth and the Sun is 3.52×10^{22} newtons. Calculate the mass of the Sun. [Show all work, including the equation and substitution with units.]

Base your answers to question 11 on the information below.

    The driver of a car made an emergency stop on a straight horizontal road. The wheels locked and the car skidded to a stop. The marks made by the rubber tires on the dry asphalt are 16 meters long, and the car's mass is 1200 kilograms.

11. Determine the weight of the car. [Show all work, including the equation and substitution with units.]

12. As an astronaut travels from the surface of Earth to a position that is four times as far away from the center of Earth, the astronaut's

    (1) mass decreases
    (2) mass remains the same
    (3) weight increases
    (4) weight remains the same   12 _____

13. A satellite weighs 200 newtons on the surface of Earth. What is its weight at a distance of one Earth radius above the surface of Earth?

    (1) 50 N   (3) 400 N
    (2) 100 N   (4) 800 N   13 _____

14. A container of rocks with a mass of 65.0 kilograms is brought back from the Moon's surface where the acceleration due to gravity is 1.62 meters per second². What is the weight of the container of rocks on Earth's surface?

    (1) 638 N   (3) 105 N
    (2) 394 N   (4) 65.0 N   14 _____

15. A 70.-kilogram astronaut has a weight of 560 newtons on the surface of planet Alpha. What is the acceleration due to gravity on planet Alpha?

    (1) 0.0 m/s²   (3) 9.8 m/s²
    (2) 8.0 m/s²   (4) 80. m/s²   15 _____

16. The graph below represents the relationship between gravitational force and mass for objects near the surface of Earth. The slope of the graph represents the

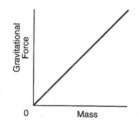

    (1) acceleration due to gravity
    (2) universal gravitational constant
    (3) momentum of objects
    (4) weight of objects   16 _____

17. An astronaut weighs 8.00×10^2 newtons on the surface of Earth. What is the weight of the astronaut 6.37×10^6 meters above the surface of Earth?

    (1) 0.00 N
    (2) 2.00×10^2 N
    (3) 1.60×10^3 N
    (4) 3.20×10^3 N   17 _____

18. A 1200-kilogram space vehicle travels at 4.8 meters per second along the level surface of Mars. If the magnitude of the gravitational field strength on the surface of Mars is 3.7 newtons per kilogram, the magnitude of the normal force acting on the vehicle is

    (1) 320 N   (3) 4400 N
    (2) 930 N   (4) 5800 N   18 _____

Base your answers to questions 19 *a* and *b* on the accompanying data table.

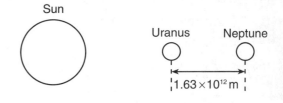

Data Table

Mass of the Sun	1.99×10^{30} kg
Mass of Uranus	8.73×10^{25} kg
Mass of Neptune	1.03×10^{26} kg
Mean distance of Uranus to the Sun	2.87×10^{12} m
Mean distance of Neptune to the Sun	4.50×10^{12} m

19. *a*) The accompanying diagram represents Neptune, Uranus, and the Sun in a straight line. Neptune is 1.63×10^{12} meters from Uranus.

Calculate the magnitude of the interplanetary force of attraction between Uranus and Neptune at this point. [Show all work, including the equation and substitution with units.]

Sun

Uranus   Neptune

1.63×10^{12} m

(Not drawn to scale)

b) The magnitude of the force the Sun exerts on Uranus is 1.41×10^{21} newtons. Explain how it is possible for the Sun to exert a greater force on Uranus than Neptune exerts on Uranus.

_____

_____

Base your answer to question 20 on the information below.

Io (pronounced "EYE oh") is one of Jupiter's moons discovered by Galileo. Io is slightly larger than Earth's Moon. The mass of Io is 8.93×10^{22} kilograms and the mass of Jupiter is 1.90×10^{27} kilograms. The distance between the centers of Io and Jupiter is 4.22×10^{8} meters.

20. Calculate the magnitude of the gravitational force of attraction that Jupiter exerts on Io. [Show all work, including the equation and substitution with units.]

Gravity
Answers
Set 1

1. 1 Under Mechanics, find the equation $g = F_g/m$. Substitution into this equation for F_g and the given value for the acceleration due to gravity on planet X gives 19.6 m/ s^2 = (980. N)/m. Solving for m yields 50.0 kg.

2. 3 The equation needed is $F_g = \dfrac{Gm_1m_2}{r^2}$. The universal gravitational constant (G) is found in the reference table on the List of Physical Constants. Substitution into the equation gives

    $$F_g = \frac{(6.67 \times 10^{-11} \text{ N} \bullet \text{m}^2/\text{kg}^2)(15.0 \text{ kg})(15.0 \text{ kg})}{(3.00 \text{ m})^2}$$

    Solving gives $F_g = 1.67 \times 10^{-9}$ N.

3. 3 Under Mechanics, find the equation $F_g = Gm_1m_2/r^2$. This equation indicates that the gravitational force between two objects varies inversely with the square of the distance between their centers. In moving from 16 Earth radii to 2 Earth radii from the center of the Earth, the distance has become 1/8 as great. Squaring 1/8 gives 1/64 and inverting shows that the force has become 64 times as great.

4. 3 In the reference table, find the equation $g = F_g/m$. The weight of the object (F_g) is 45 N. The mass is 15 kg. Substitution gives g = 45 N/15 kg. Solving, g = 3.0 m/ s^2.

5. 1 The force of the table on the block must be equal and opposite the force of the block on the table, which is the weight of the block. Find the equation $g = F_g/m$. The value of g is on the List of Physical Constants. Substituting into the equation gives 9.81 m/ s^2 = F_g/(5.00 kg). Solving for F_g yields a weight of 49.1 N. The force of the table on the block is therefore 49.1 N upward.

6. 4 Since the skydiver is falling at a constant speed, the skydiver is in a state of equilibrium. This means that the net fore acting on the skydiver is zero. The upward force of air friction acting on the skydiver must then be equal to the weight of the skydiver. Using the equation $g = F_g/m$. Substitution gives 9.81 m/ s^2 = $\dfrac{F_g}{60 \text{ kg}}$ and solving yields F_g = 589 N or 600 N.

7. 3 This is an application of Newton's Third Law which states that for every action, there is an equal but opposite reaction. Let the action be the gravitational force of the Earth on the Moon. The reaction is the equal but opposite gravitational force of the Moon on the Earth.

8. 3   The gravitational force between two objects is one of attraction. The Earth and satellite attract each other with equal but opposite forces. The force of the Earth on the satellite is directed toward the Earth and the force of the satellite on the Earth is directed toward the satellite. Choice 3 shows the proper directions for these forces.

9. 1   Newton's Third Law states that for every action, there is an equal but opposite reaction. If the Earth exerts a gravitational force of magnitude F on Moon, the Moon must exert a gravitational force of magnitude F on the Earth.

10. $F = G\dfrac{m_1 m_2}{r^2}$

       $m_2 = \dfrac{Fr^2}{Gm_1}$

       $m_2 = \dfrac{\left(3.52 \times 10^{22}\,\text{N}\right)\left(1.50 \times 10^{11}\,\text{m}\right)^2}{\left(6.67 \times 10^{-11}\,\dfrac{\text{N} \bullet \text{m}^2}{\text{kg}^2}\right)\left(5.98 \times 10^{24}\,\text{kg}\right)}$

       $m_2 = 1.99 \times 10^{30}\,\text{kg}$

Explanation: In the equation for gravitational force of attraction between two masses, one mass is that of the Earth (m_1) and the other mass is the mass of the Sun (m_2). The mass of the Earth, the universal gravitational constant (G), and the distance between the Earth and Sun are found on the List of Physical Constants. Substitute the values into the equation and calculate the mass of the Sun.

11. $g = F_g/m$
       $F_g = mg$
       $F_g = (1200\ \text{kg})(9.81\ \text{m/s}^2)$
       $F_g = 11.800\ \text{N}$ *or* $12.00\ \text{N}$

Explanation: Under Mechanics, find the equation $g = F_g/m$. The value of g is found on the List of Physical Constants. Substitute into the equation and solve for the weight of the car.

Mechanics – Friction

Friction: Friction is a force that always opposes motion. The force of friction is the product of the coefficient of friction, a constant that depends upon the materials and their surfaces, and the normal force pressing the surfaces together.

$$F_f = \mu F_N$$

where: F_f = force of friction
F_N = normal force
μ = coefficient of friction

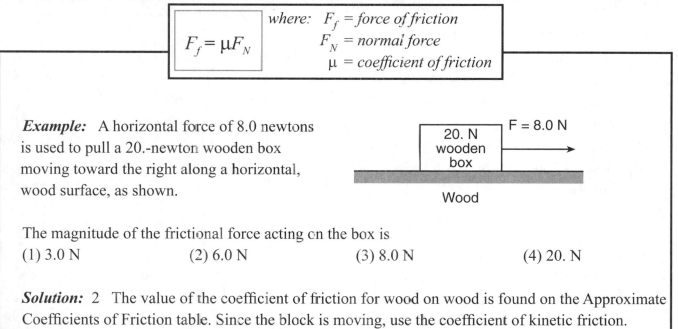

Example: A horizontal force of 8.0 newtons is used to pull a 20.-newton wooden box moving toward the right along a horizontal, wood surface, as shown.

The magnitude of the frictional force acting on the box is
(1) 3.0 N   (2) 6.0 N   (3) 8.0 N   (4) 20. N

Solution: 2   The value of the coefficient of friction for wood on wood is found on the Approximate Coefficients of Friction table. Since the block is moving, use the coefficient of kinetic friction. Substituting into the equation gives $F_f = (0.30)(20.\text{ N})$. Solving, $F_f = 6.0$ N.

Friction - Additional Information:

• Coefficients of friction are found on the Approximate Coefficients of Friction table.

• The normal force (F_N) is the force acting perpendicular to the surfaces of the object. On a horizontal surface, it is the weight of the object. On an incline, it is the component of the objects weight that is perpendicular to the incline.

• Be sure to read the question carefully to understand whether the coefficient of static friction or the coefficient of kinetic friction should be used.

  Use the coefficient of static friction when starting an object in motion.

  Use the coefficient of kinetic friction when the object is in motion.

• When work is done only against friction, the internal energy of the object increases.

1. The force required to start an object sliding across a uniform horizontal surface is larger than the force required to keep the object sliding at a constant velocity. The magnitudes of the required forces are different in these situations because the force of kinetic friction

    (1) increases as the speed of the object relative to the surface increases
    (2) decreases as the speed of the object relative to the surface increases
    (3) is greater than the force of static friction
    (4) is less than the force of static friction   1 _____

2. A car's performance is tested on various horizontal road surfaces. The brakes are applied, causing the rubber tires of the car to slide along the road without rolling. The tires encounter the greatest force of friction to stop the car on

    (1) dry concrete   (3) wet concrete
    (2) dry asphalt   (4) wet asphalt   2 _____

3. The diagram below shows a granite block being slid at constant speed across a horizontal concrete floor by a force parallel to the floor.

    Which pair of quantities could be used to determine the coefficient of friction for the granite on the concrete?

    (1) mass and speed of the block
    (2) mass and normal force on the block
    (3) frictional force and speed of the block
    (4) frictional force and normal force on the block   3 _____

4. What is the magnitude of the force needed to keep a 60.-newton rubber block moving across level, dry asphalt in a straight line at a constant speed of 2.0 meters per second?

    (1) 40. N   (3) 60. N
    (2) 51 N   (4) 120 N   4 _____

Base your answers to question 5 on the information below.

A force of 10. newtons toward the right is exerted on a wooden crate initially moving to the right on a horizontal wooden floor. The crate weighs 25 newtons.

5. Calculate the magnitude of the force of friction between the crate and the floor. [Show all work, including the equation and substitution with units.]

Base your answers to questions 6 *a* and *b* on the information below.

The driver of a car made an emergency stop on a straight horizontal road. The wheels locked and the car skidded to a stop. The marks made by the rubber tires on the dry asphalt are 16 meters long, and the car's mass is 1200 kilograms.

6. *a*) Determine the weight of the car.

b) Calculate the magnitude of the frictional force the road applied to the car in stopping it. [Show all work, including the equation and substitution with units.]

7. A 10.-kilogram rubber block is pulled horizontally at constant velocity across a sheet of ice. Calculate the magnitude of the force of friction acting on the block. [Show all work, including the equation and substitution with units.]

8. When a 12-newton horizontal force is applied to a box on a horizontal tabletop, the box remains at rest. The force of static friction acting on the box is

   (1) 0 N
   (2) between 0 N and 12 N
   (3) 12 N
   (4) greater than 12 N   8 _____

9. An 80-kilogram skier slides on waxed skis along a horizontal surface of snow at constant velocity while pushing with his poles. What is the horizontal component of the force pushing him forward?

   (1) 0.05 N   (3) 40 N
   (2) 0.4 N   (4) 4 N   9 _____

Note: Question 10 has only three choices.

10. Compared to the force needed to start sliding a crate across a rough level floor, the force needed to keep it sliding once it is moving is

    (1) less
    (2) greater
    (3) the same   10 _____

11. A box is pushed toward the right across a classroom floor. The force of friction on the box is directed toward the

    (1) left   (3) ceiling
    (2) right   (4) floor   11 _____

12. When a force moves an object over a rough, horizontal surface at a constant velocity, the work done against friction produces an increase in the object's

    (1) weight   (3) potential energy
    (2) momentum   (4) internal energy   12 _____

Note: Question 13 has only three choices.

13. The accompanying diagram shows a block sliding down a plane inclined at angle θ with the horizontal. As angle θ is increased, the coefficient of kinetic friction between the bottom surface of the block and the surface of the incline will

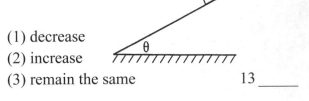

    (1) decrease
    (2) increase
    (3) remain the same   13 _____

14. What is the magnitude of the force needed to keep a 50.-newton rubber block moving across level, dry concrete in a straight line at a constant speed of 2.0 meters per second?

    (1) 34 N   (3) 60. N
    (2) 45 N   (4) 120 N   14 _____

15. The table below lists the coefficients of kinetic friction for four materials sliding over steel.

Material	Coefficient of Kinetic Friction
aluminum	0.47
brass	0.44
copper	0.36
steel	0.57

A 10.-kilogram block of each of these materials is pulled horizontally across a steel floor at constant velocity. Which block requires the smallest applied force to keep it moving at constant velocity?

    (1) aluminum   (3) copper
    (2) brass   (4) steel   15 _____

16. Explain how to find the coefficient of kinetic friction between a wooden block of unknown mass and a tabletop in the laboratory. Include the following in your explanation:

- Measurements required
- Procedure
- Equipment needed
- Equation(s) needed to calculate the coefficient of friction

Base your answer to question 17 on the information below.

A 1200-kilogram car moving at 12 meters per second collides with a 2300-kilogram car that is waiting at rest at a traffic light. After the collision, the cars lock together and slide. Eventually, the combined cars are brought to rest by a force of kinetic friction as the rubber tires slide across the dry, level, asphalt road surface.

17. Calculate the magnitude of the frictional force that brings the locked-together cars to rest. [Show all work, including the equation and substitution with units.]

Base your answers to questions 18a and b on the information and the accompanying diagram.

A 10.-kilogram box, sliding to the right across a rough horizontal floor, accelerates at −2.0 meters per second² due to the force of friction.

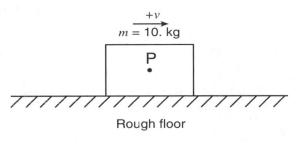

Rough floor

18. a) Calculate the magnitude of the net force acting on the box. [Show all work, including the equation and substitution with units.]

b) Calculate the coefficient of kinetic friction between the box and the floor. [Show all work, including the equation and substitution with units.]

1. **4** Under Mechanics, find the equation $F_f = \mu F_N$. For a given object, the force of friction depends upon the coefficient of friction (μ). On the table of Approximate Coefficients of Friction, the coefficients of kinetic friction are all less than the coefficients of static friction. Hence the force of kinetic friction is less than the force of static friction.

2. **1** Locate the equation $F_f = \mu F_N$. The normal force is the weight of the car, which is constant. The frictional force varies directly with the coefficient of friction. Since the tires are sliding, the coefficient of kinetic friction must be used. Using the table of Approximate Coefficients of Friction, the largest value is that for rubber and dry concrete.

3. **4** The coefficient of friction (μ) is defined as the ratio of the frictional force to the normal force pressing the surfaces together $\mu = F_f / F_N$.

4. **1** Find the equation $F_f = \mu F_N$ in the reference table. The normal force is the weight of the block (60. N). Since the block is moving or sliding along the surface, the coefficient of kinetic friction between rubber and dry asphalt must be used (see the table of Approximate Coefficients of Friction). Substituting into the equation and solving, $F_f = (0.67)(60.\text{ N}) = 40.0\text{ N}$.

5. $F_f = \mu F_N$
   $F_f = (0.30)(25\text{ N})$
   $F_f = 7.5\text{ N}$

   Explanation: Under Mechanics, find the equation $F_f = \mu F_N$. On a horizontal surface, the weight of the object is the normal force. Find the coefficient of kinetic friction for Wood on wood in the table of Approximate Coefficients of Friction. Substitute into the equation and solve for F_f.

6. a) $g = F_g / m$
   $F_g = mg$
   $F_g = (1200\text{ kg})(9.81\text{ m/s}^2)$
   $F_g = 11,800\text{ N}$ *or* $12,000\text{ N}$

   Explanation: Locate the equation $g = F_g / m$. The value of g is found on the List of Physical Constants. Substitute into the equation and solve of the weight of the car.

   b) $F_f = \mu F_N$
   $F_f = (0.67)(12,000\text{ N})$
   $F_f = 8,000\text{ N}$ *or* $8,040\text{ N}$

   Explanation: Use the equation $F_f = \mu F_N$. The normal force is the weight of the car (your answer to 6a) since the car is on a horizontal road. The coefficient of friction between rubber and dry asphalt is found on the table of Approximate Coefficients of Friction. Substitute into the equation and solve for F_f.

7. $F_f = \mu F_N$
   $F_f = (1.5)(10\text{ kg})(9.81\text{ m/s}^2)$
   $F_f = 15\text{ N}$ *or* 14.7 N

   The normal force is the weight of the block of ice. Under Mechanics, find the equation $g = F_g / m$. Solving for F_g gives $F_g = mg$. The value of μ is found on the Approximate Coefficients of Friction table and g is found in the List of Physical Constants. Substitute the values into the equation and solve for F_f.

Mechanics – Momentum and Impulse

Momentum and Impulse: Momentum is defined as a product of the mass of an object and its velocity. Impulse is equal to the change in momentum an object undergoes.

Momentum is defined as the product of the mass of an object and its velocity.

$$p = mv$$

where: p = momentum
m = mass
v = velocity or speed

Example: What is the speed of a 1.0×10^3-kilogram car that has a momentum of 2.0×10^4 kilogram•meters per second east?

(1) 5.0×10^{-2} m/s (2) 2.0×10^1 m/s (3) 1.0×10^4 m/s (4) 2.0×10^7 m/s

Solution: 2 Speed is the magnitude of the velocity vector, which is toward the east. Substitution into the equation $p = mv$ gives 2.0×10^4 kg•m/s = 1.0×10^3 kg(v). Solving, $v = 2.0 \times 10^1$ m/s.

Momentum is a quantity that is conserved, meaning that it does not change as a result of an interaction or collision between two objects. Momentum may be transferred from one object to another during an interaction but does not change as a result of that interaction. This equation expresses the Law of Conservation of Momentum.

$$P_{before} = P_{after}$$

where: p = momentum

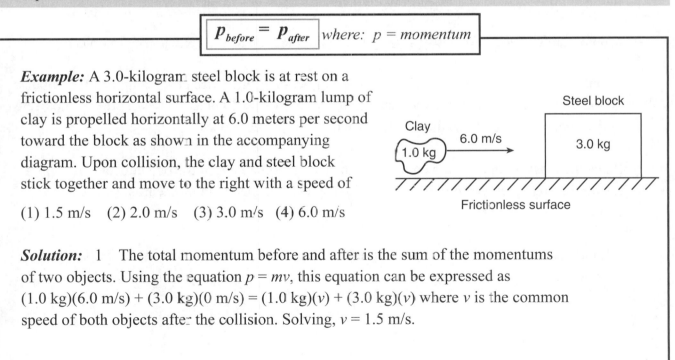

Example: A 3.0-kilogram steel block is at rest on a frictionless horizontal surface. A 1.0-kilogram lump of clay is propelled horizontally at 6.0 meters per second toward the block as shown in the accompanying diagram. Upon collision, the clay and steel block stick together and move to the right with a speed of

(1) 1.5 m/s (2) 2.0 m/s (3) 3.0 m/s (4) 6.0 m/s

Solution: 1 The total momentum before and after is the sum of the momentums of two objects. Using the equation $p = mv$, this equation can be expressed as (1.0 kg)(6.0 m/s) + (3.0 kg)(0 m/s) = (1.0 kg)(v) + (3.0 kg)(v) where v is the common speed of both objects after the collision. Solving, $v = 1.5$ m/s.

Impulse is defined as the product of the net or unbalanced force acting on an object and the time that the force acts on the object. It is also equal to the change in momentum of the object.

$$J = F_{net}\, t = \Delta p$$

where: J = impulse
F_{net} = net force
t = time
Δp = change in momentum

Example: A 60-kilogram student jumps down from a laboratory counter. At the instant he lands on the floor his speed is 3 meters per second. If the student stops in 0.2 second, what is the average force of the floor on the student?

(1) 1×10^{-2} N (2) 1×10^2 N (3) 9×10^2 N (4) 4 N

Solution: 3 The change in the students momentum (Δp) can be expressed as $m\Delta v$, where Δv is the change in the students speed (the object starts from rest and reaches a speed of 3 m/s). Therefore $F_{net}\, t = m\Delta v$. Substituting into this equation gives $F(0.2 \text{ s}) = (60 \text{ kg})(3 \text{ m/s})$. Solving, $F = 9 \times 10^2$ N.

Momentum and Impulse - Additional Information:

- Whenever a net force acts on an object and changes its velocity, it produces a change in momentum or delivers an impulse to the object.

- Momentum is a vector quantity, having the same direction as the velocity of the object.

- Impulse is a vector quantity, having the same direction as the force producing it. If the impulse is in the same direction as the momentum, the velocity of the object increases. If the impulse is in the opposite direction as the momentum, the velocity decreases.

- Since the mass of an object is constant, a change in momentum of the object occurs only as a result of a change in velocity of the object. Therefore a change in momentum (Δp) can be expressed as $\Delta p = m\Delta v$.

- Momentum is a quantity that is conserved, meaning that it is the same after as before an interaction.

1. Which two quantities can be expressed using the same units?

   (1) energy and force
   (2) momentum and energy
   (3) impulse and force
   (4) impulse and momentum   1 _____

2. At the circus, a 100.-kilogram clown is fired at 15 meters per second from a 500.-kilogram cannon. What is the recoil speed of the cannon?

   (1) 75 m/s   (3) 3.0 m/s
   (2) 15 m/s   (4) 5.0 m/s   2 _____

3. Ball A of mass 5.0 kilograms moving at 20. meters per second collides with ball B of unknown mass moving at 10. meters per second in the same direction. After the collision, ball A moves at 10. meters per second and ball B at 15 meters per second, both still in the same direction. What is the mass of ball B?

   (1) 6.0 kg   (3) 10. kg
   (2) 2.0 kg   (4) 12 kg   3 _____

4. A 50.-kilogram student threw a 0.40-kilogram ball with a speed of 20. meters per second. What was the magnitude of the impulse that the student exerted on the ball?

   (1) 8.0 N•s   (3) 4.0×10^2 N•s
   (2) 78 N•s   (4) 1.0×10^3 N•s   4 _____

5. The accompanying diagram represents two masses before and after they collide. Before the collision, mass m_A is moving to the right with speed v, and mass m_B is at rest. Upon collision, the two masses stick together.

   Before Collision   **After Collision**

   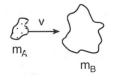

   Which expression represents the speed, v', of the masses after the collision? [Assume no outside forces are acting on m_A or m_B.]

   (1) $\dfrac{m_A + m_B v}{m_A}$ (3) $\dfrac{m_B v}{m_A + m_B}$

   (2) $\dfrac{m_A + m_B}{m_A v}$ (4) $\dfrac{m_A v}{m_A + m_B}$ 5 _____

6. The diagram below shows a 4.0-kilogram cart moving to the right and a 6.0-kilogram cart moving to the left on a horizontal frictionless surface.

   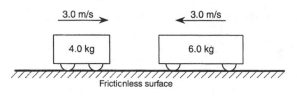

   When the two carts collide they lock together. The magnitude of the total momentum of the two-cart system after the collision is

   (1) 0.0 kg • m/s
   (2) 6.0 kg • m/s
   (3) 15 kg • m/s
   (4) 30. kg • m/s   6 _____

7. A 6.0-kilogram block, sliding to the east across a horizontal, frictionless surface with a momentum of 30. kilogram•meters per second, strikes an obstacle. The obstacle exerts an impulse of 10. newton•seconds to the west on the block. The speed of the block after the collision is

(1) 1.7 m/s   (2) 3.3 m/s   (3) 5.0 m/s   (4) 20. m/s   7_____

Base your answer to question 8 on the information below.

A 50.-kilogram child running at 6.0 meters per second jumps onto a stationary 10.-kilogram sled. The sled is on a level frictionless surface.

8. Calculate the speed of the sled with the child after she jumps onto the sled. [Show all work, including the equation and substitution with units.]

9. A 1000-kilogram car traveling due east at 15 meters per second is hit from behind and receives a forward impulse of 6000 newton-seconds. Determine the magnitude of the car's change in momentum due to this impulse.

Mechanics

10. In the diagram below, scaled vectors represent the momentum of each of two masses, *A* and *B*, sliding toward each other on a frictionless, horizontal surface.

Mass A   Frictionless surface   Mass B

Which scaled vector best represents the momentum of the system after the masses collide?

(1) ←—

(2) ——————→

(3) ←——————

(4) ——————————→   10 ____

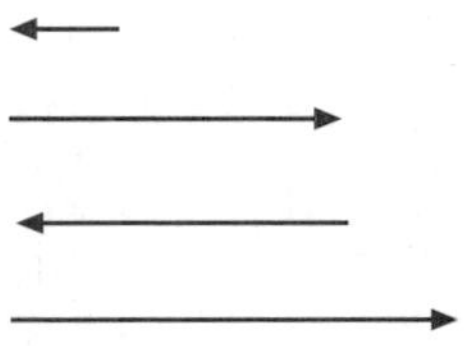

11. Cart *A* has a mass of 2 kilograms and a speed of 3 meters per second. Cart *B* has a mass of 3 kilograms and a speed of 2 meters per second. Compared to the inertia and magnitude of momentum of cart *A*, cart *B* has

(1) the same inertia and a smaller magnitude of momentum
(2) the same inertia and the same magnitude of momentum
(3) greater inertia and a smaller magnitude of momentum
(4) greater inertia and the same magnitude of momentum   11 ___

12. Which is an acceptable unit for impulse?

(1) N • m   (3) J • s
(2) J/s   (4) kg • m/s   12 ____

13. Which situation will produce the greatest change of momentum for a 1.0-kilogram cart?

(1) accelerating it from rest to 3.0 m/s
(2) accelerating it from 2.0 m/s to 4.0 m/s
(3) applying a net force of 5.0 N for 2.0 s
(4) applying a net force of 10.0 N for 0.5 s

13 ____

14. A 1.0-kilogram laboratory cart moving with a velocity of 0.50 meter per second due east collides with and sticks to a similar cart initially at rest. After the collision, the two carts move off together with a velocity of 0.25 meter per second due east. The total momentum of this frictionless system is

(1) zero before the collision
(2) zero after the collision
(3) the same before and after the collision
(4) greater before the collision than after the collision   14 ____

15. Calculate the magnitude of the impulse applied to a 0.75-kilogram cart to change its velocity from 0.50 meter per second east to 2.00 meters per second east. [Show all work, including the equation and substitution with units.]

Base your answer to question 16 on the information below.

A 1200-kilogram car moving at 12 meters per second collides with a 2300-kilogram car that is waiting at rest at a traffic light. After the collision, the cars lock together and slide. Eventually, the combined cars are brought to rest by a force of kinetic friction as the rubber tires slide across the dry, level, asphalt road surface.

16. Calculate the speed of the locked-together cars immediately after the collision. [Show all work, including the equation and substitution with units.]

Base your answers to question 17 on the information and diagram below.

A 1000.-kilogram empty cart moving with a speed of 6.0 meters per second is about to collide with a stationary loaded cart having a total mass of 5000. kilograms, as shown. After the collision, the carts lock and move together. [Assume friction is negligible.]

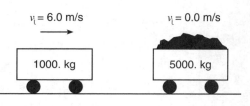

17. Calculate the speed of the combined carts after the collision. [Show all work, including the equation and substitution with units.]

18. A motorcycle being driven on a dirt path hits a rock. Its 60.-kilogram cyclist is projected over the handlebars at 20. meters per second into a haystack. If the cyclist is brought to rest in 0.50 second, what is the magnitude of the average force exerted on the cyclist by the haystack? [Show all work, including the equation and substitution with units.]

Copyright © 2011
Topical Review Book Company

1. **4**   Under Mechanics, find the equation $J = F_{net}\,t = \Delta p$. Since J and Δp are equal to each other, they must have the same units.

2. **3**   This problem deals with the Law of Conservation of Momentum. In the reference table, find the equations $p = mv$ and $p_{before} = p_{after}$. Since the clown and the cannon are both at rest before the cannon is fired, the $p_{before} = 0$. The total momentum after the cannon is fired is the sum of the momentum of the clown, $p = (100.\text{ kg})(15\text{ m/s})$ and the momentum of the cannon, $p = (500.\text{ kg})(v)$. Substituting into the conservation equation gives $0 = (100.\text{ kg})(15\text{ m/s}) + (500\text{ kg})(v)$. Solving gives $v = -3.0$ m/s. The negative sign indicates that the cannon recoils in the opposite direction to that of the motion of the clown.

3. **3**   In the reference table, locate the equations $p_{before} = p_{after}$ and $p = mv$. The total momentum before and after the collision is the sum of the momentum of ball A and ball B. Thus, $(5.0\text{ kg})(20.\text{ m/s}) + m_B (10.\text{ m/s}) = (5.0\text{ kg})(10.\text{ m/s}) + m_B(15\text{ m/s})$. Solving for m_B gives $50.\text{ kg}\cdot\text{m/s} = (5.0\text{ m/s})(m_B)$. Then $m_B = 10.$ kg.

4. **1**   Find the equations $J = \Delta p$ and $p = mv$. A change in momentum of an object is produced by a change in the objects velocity. Therefore $\Delta p = m\Delta v$. Using the balls mass and a change in velocity of 20. m/s gives $\Delta p = (0.40\text{ kg})(20.\text{ m/s}) = 8.0$ N•s. This is the impulse delivered to the ball. The mass of the student has nothing to do with the solution to this problem.

5. **4**   The equations needed are $p = mv$ and $p_{before} = p_{after}$. Before the collision, the momentum of mass m_A is $m_A v$ and that of mass m_B is 0 since it is at rest. After the collision, the momentum of mass m_A is $m_A v'$ and that of mass m_B is $m_B v'$. Substitution into the second equation gives $m_A v + 0 = m_A v' + m_B v'$. Solving, $v' = (m_A v)/(m_A + m_B)$.

6. **2**   The equations needed are $p = mv$ and $p_{before} = p_{after}$. The total momentum before the collision is the sum of the momentum of both carts. The momentum of the 4.0 kg cart is $(4.0\text{ kg})(3.0\text{ m/s}) = 12\text{ kg•m/s}$. The momentum of the 6.0 kg cart is $(6.0\text{ kg})(3.0\text{ m/s}) = 18$ kg•m/s. Because momentum is a vector quantity, the momentum of the 6.0 kg cart is negative to show an opposite direction to that of the 4.0 kg cart. Therefore, $p_{before} = 12\text{ kg•m/s} + (-18\text{ kg•m/s}) = -6.0$ kg•m/s, in the direction of motion of the 6.0 kg cart. Since only the magnitude of the momentum after the collision is needed, the negative sign is not included in the answer.

7. 2   Under Mechanics, find the equation $J = F_{net}\, t = \Delta p$. The change in momentum of an object equals the impulse received. Both momentum and impulse are vector quantities. An impulse directed toward the west received by an object with a momentum directed toward the east decreases the momentum of the object. Therefore, after receiving the impulse, the momentum of the object is 30. kg•m/s – 10. kg•m/s = 20. kg•m/s. Now find the equation $p = mv$ under Mechanics. Substituting into the equation gives 20. kg•m/s = (6.0 kg)v. Solving, $v = 3.3$ m/s.

8. $P_{before} = P_{after}$

    or

$m_{before}\, v_{before} = m_{after}\, v_{after}$
(50. kg)(6.0 m/s) = (60. kg) v_{after}
v_{after} = (50. kg)(6.0 m/s) / (60. kg)
v_{after} = 5.0 m/s

    or

$P_{before} = P_{after}$
$m_{child} + m_{sled} = m_{child} + m_{sled}$
(50. kg)(6.0 m/s) + 0 = (50. kg)(v_{after}) + (10. kg)(v_{after})
(50. kg)(6.0 m/s) + 0 = (50. kg + 10. kg)(v_{after})
(50. kg)(6.0 m/s) + 0 = (60. kg)(v_{after})
v_{after} = (50. kg)(6.0 m/s) / (6.0 kg)
v_{after} = 5.0 m/s

Explanation: Under Mechanics, find the equations $p_{before} = p_{after}$ (Law of Conservation of Momentum) and $p = mv$. The momentum of the sled before is zero since it is at rest. The combined mass *after* is the mass of the boy and sled and they travel with a common velocity. Substitution gives (50. kg)(6.0 m/s) = (60 kg)v_{after}. Solve for v_{after}.

    or

Using the same equations, express the momentum of the child and sled separately before and after and solve for v_{after}.

9. Answer 6000 N•s   *or*   6000 kg•m/s.

Explanation: In the reference table, find the equation $J = F_{net}\, t = \Delta p$. This equation tells us that the change in momentum is equal to the impulse received, which is 6000 N•s.

Springs and Elastic Potential Energy: An applied force is necessary to stretch or compress a spring. The magnitude of this force is equal to the product of the spring constant and the change in length of the spring from its equilibrium or natural position.

$$F_s = kx$$

where: F_s = force on a spring
k = spring constant
x = change in spring length from the equilbrium position

Example: The spring in a scale in the produce department of a supermarket stretches 0.025 meter when a watermelon weighing 1.0×10^2 newtons is placed on the scale. The spring constant for this spring is

(1) 3.2×10^5 N/m   (2) 4.0×10^3 N/m   (3) 2.5 N/m   (4) 3.1×10^{-2} N/m

Solution: 2 The force causing the spring to stretch is the weight of the watermelon. Substitution into the equation gives 1.0×10^2 N $= k(0.025$ m$)$. Solving gives $k = 4.0 \times 10^3$ N/m.

When a force is used to stretch or compress a spring, work is done on the spring and is stored in the spring as elastic potential energy. The amount of this energy is equal to one-half the product of the spring constant and the square of the change in length of the spring.

$$PE_s = \frac{1}{2}kx^2$$

where: PE_s = potential energy stored in a spring
k = spring constant
x = change in spring length from the equilbrium position

Example: A catapult with a spring constant of 1.0×10^4 newtons per meter is required to launch an airplane from the deck of an aircraft carrier. The plane is released when it has been displaced 0.50 meter from its equilibrium position by the catapult. The energy acquired by the airplane from the catapult during takeoff is approximately

(1) 1.3×10^3 J   (2) 2.0×10^4 J   (3) 2.5×10^3 J   (4) 1.0×10^4 J

Solution: 1 The airplane gets its energy from the compressed spring of the catapult. Substitution into the equation gives $PE_s = \frac{1}{2}(1.0 \times 10^4$ N/m$)(0.50$ m$)^2$. Solving gives 1.3×10^3 J.

Springs and Elastic Potential Energy - Additional Information:

- As the amount of stretching or compression of a spring increases, the force required increases.

- The work done and the potential energy stored in the spring depends upon the average force.

- The spring constant (k) depends only upon the spring.

Set 1 — Springs and Elastic Potential Energy

1. Which graph best represents the elastic potential energy stored in a spring (PE_s) as a function of its elongation, x?

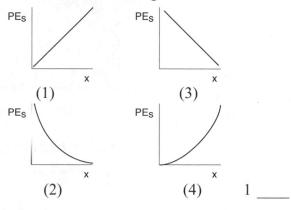

(1)

(3)

(2)

(4)

1 _____

2. The diagram represents a spring hanging vertically that stretches 0.075 meter when a 5.0-newton block is attached. The spring-block system is at rest in the position shown. The value of the spring constant is

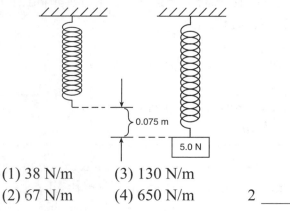

(1) 38 N/m   (3) 130 N/m
(2) 67 N/m   (4) 650 N/m   2 _____

3. When a 1.53-kilogram mass is placed on a spring with a spring constant of 30.0 newtons per meter, the spring is compressed 0.500 meter. How much energy is stored in the spring?

(1) 3.75 J   (3) 15.0 J
(2) 7.50 J   (4) 30.0 J   3 _____

Note: Question 4 has only three choices.

4. A mass, M, is hung from a spring and reaches equilibrium at position B. The mass is then raised to position A and released. The mass oscillates between positions A and C. [Neglect friction.]

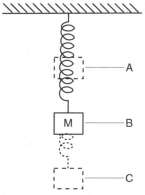

At which position is mass M located when the elastic potential energy of the system is at a maximum?

(1) A (2) B (3) C   4 _____

5. As shown in the diagram below, a 0.50-meter-long spring is stretched from its equilibrium position to a length of 1.00 meter by a weight.

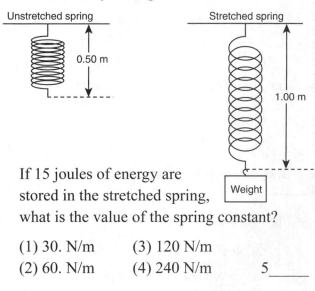

If 15 joules of energy are stored in the stretched spring, what is the value of the spring constant?

(1) 30. N/m   (3) 120 N/m
(2) 60. N/m   (4) 240 N/m   5 _____

Base your answers to questions 6 *a*, *b*, *c*, and *d* on the information and data table below.

The spring in a dart launcher has a spring constant of 140 newtons per meter. The launcher has six power settings, 0 through 5, with each successive setting having a spring compression 0.020 meter beyond the previous setting. During testing, the launcher is aligned to the vertical, the spring is compressed, and a dart is fired upward. The maximum vertical displacement of the dart in each test trial is measured. The results of the testing are shown in the accompanying table.

Directions: Using the information in the accompanying data table, construct a graph on the grid below, following the directions below.

6. *a)* Plot the data points for the dart's maximum vertical displacement versus spring compression.

b) Draw the line or curve of best fit.

c) Using information from your graph, calculate the energy provided by the compressed spring that causes the dart to achieve a maximum vertical displacement of 3.50 meters. [Show all work, including the equation and substitution with units.]

d) Determine the magnitude of the force, in newtons, needed to compress the spring 0.040 meter.

_____ N

Data Table

Power Setting	Spring Compression (m)	Dart's Maximum Vertical Displacement (m)
0	0.000	0.00
1	0.020	0.29
2	0.040	1.14
3	0.060	2.57
4	0.080	4.57
5	0.100	7.10

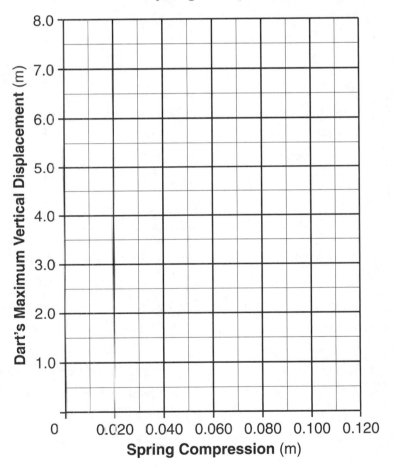

Dart's Maximum Vertical Displacement vs. Spring Compression

Note: Question 7 has only three choices.

7. The accompaning graph shows elongation as a function of the applied force for two springs, *A* and *B*. Compared to the spring constant for spring *A*, the spring constant for spring *B* is

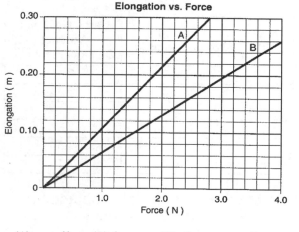

Elongation vs. Force

(1) smaller (2) larger (3) the same 7 _____

8. A 10.-newton force is required to hold a stretched spring 0.20 meter from its rest position. What is the potential energy stored in the stretched spring?

(1) 1.0 J   (3) 5.0 J
(2) 2.0 J   (4) 50. J   8 _____

9. The diagram below shows a toy cart possessing 16 joules of kinetic energy traveling on a frictionless, horizontal surface toward a horizontal spring.

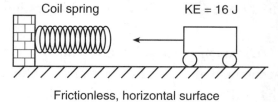

Coil spring   KE = 16 J

Frictionless, horizontal surface

If the cart comes to rest after compressing the spring a distance of 1.0 meter, what is the spring constant of the spring?

(1) 32 N/m   (3) 8.0 N/m
(2) 16 N/m   (4) 4.0 N/m   9 _____

10. A 5-newton force causes a spring to stretch 0.2 meter. What is the potential energy stored in the stretched spring?

(1) 1 J   (3) 0.2 J
(2) 0.5 J   (4) 0.1 J   10 _____

Base your answer to question 11 on the information and diagram below.

A pop-up toy has a mass of 0.020 kilogram and a spring constant of 150 newtons per meter. A force is applied to the toy to compress the spring 0.050 meter.

11. Calculate the potential energy stored in the compressed spring. [Show all work, including the equation and substitution with units.]

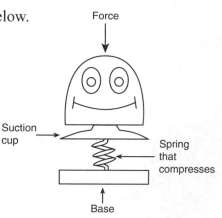

Force

Suction cup

Spring that compresses

Base

Springs and Elastic Potential Energy
Answers
Set 1

1. 4   The equation $PE_s = \frac{1}{2}kx^2$ indicates that elastic potential energy varies directly with the square of the elongation. Graph 4 shows a direct square relationship. As the elongation increases, the elastic potential energy increases at a more rapid rate.

2. 2   Under Mechanics, find the equation $F_s = kx$. Substitution gives $5.0 \text{ N} = k(0.075 \text{ m})$. Solving, $k = 67 \text{ N/m}$.

3. 1   In the reference table, locate the equation $PE_s = \frac{1}{2}kx^2$. Substitution into this equation gives $PE_s = \frac{1}{2}(30.0 \text{ N/m})(0.500 \text{ m})^2$ and $PE_s = 3.75 \text{ J}$. The mass of the object is not a factor in this calculation.

4. 3   Under Mechanics, locate the equation $PE_s = \frac{1}{2}kx^2$. Since k is a constant for the spring, the only variable determining the elastic potential energy of the spring is the elongation. The spring has its greatest elongation at point C and therefore has its maximum potential energy at point C.

5. 3   The equation needed for this problem is $PE_s = \frac{1}{2}kx^2$. The change in length of the spring is $1.00 \text{ m} - 0.50 \text{ m} = 0.50 \text{ m}$. Substitution into the equation gives $15 \text{ J} = \frac{1}{2}k(0.50 \text{ m})^2$. Solving, $k = 120 \text{ N/m}$.

6. *a-b*

Dart's Maximum Vertical Displacement vs. Spring Compression

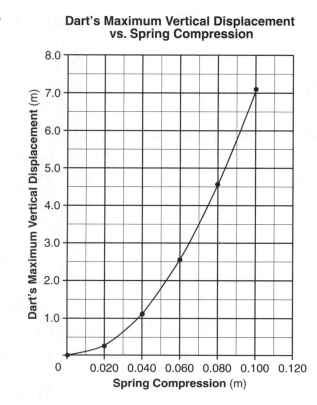

a) Explanation: Each point must be plotted to $\pm$ 0.3 grid space.

b) Explanation: Connect the points plotted on the graph with a smooth curve for best fit.

c)
$$PE_s = \frac{1}{2}kx^2$$
$$PE_s = \frac{1}{2}(140 \text{ N/m})(0.070 \text{ m})^2$$
$$PE_s = 0.34 \text{ J}$$

Explanation: Under Mechanics, find the equation $PE_s = \frac{1}{2}kx^2$. Using the graph, to achieve a maximum vertical displacement of 3.5 m, the spring must be compressed 0.07 m. Substituting and solving, $PE_s = \frac{1}{2}(140 \text{ N/m})(0.07 \text{ m})^2 = 0.34 \text{ J}$.

d) Answer: 5.6 N

Explanation: To solve this problem, one must use the equation $F_s = kx$. Substituting and solving, $F_s = (140 \text{ N/m})(0.04 \text{ m}) = 5.6 \text{ N}$.

Work, Energy and Power: Work is done when a force acts on matter and moves it against some opposition, such as friction or gravity. Energy may be defined as a measure of the ability to do work. Power is a measure of the rate at which work is done or energy is transferred.

Potential energy is the energy possessed by an object due to its position or condition. This equation is used to calculate the gravitational potential energy of an object in the gravitational field of the Earth. It is measured relative to the Earth's surface, where the gravitational potential energy is defined to be zero. The Δh term is the height of the object above the Earth's surface.

$$\Delta PE = mg\Delta h$$

where: ΔPE = change in potential energy
m = mass
g = acceleration due to gravity or gravitational field strength
Δh = change in height

Example: A 60.-kilogram student climbs a ladder a vertical distance of 4.0 meters in 8.0 seconds. Approximately how much total work is done against gravity by the student during the climb?

(1) 2.4×10^3 J   (2) 2.9×10^2 J   (3) 2.4×10^2 J   (4) 3.0×10^1 J

Solution: 1 The work done against gravity by the student will be equal to the increase in potential energy as the student climbs the ladder. The value of g is found on the List of Physical Constants. Substitution gives $\Delta PE = W = (60.$ kg$)(9.81$ m/s$^2)(4.0$ m$)$. Solving gives 2.4×10^3 J. The time has no effect on the work done by the student.

Kinetic energy is the energy an object possesses due to its motion. It is equal to one-half the product of the mass of an object and the square of its speed. Since kinetic energy varies directly with the speed, the kinetic energy of an object increases rapidly as the speed increases.

$$KE = \tfrac{1}{2}mv^2$$

where: KE = kinetic energy
m = mass
v = velocity or speed

Example: A 45.0-kilogram boy is riding a 15.0-kilogram bicycle with a speed of 8.00 meters per second. What is the combined kinetic energy of the boy and the bicycle?

(1) 240. J   (2) 480. J   (3) 1440 J   (4) 1920 J

Solution: 4 The mass is the combined mass of the boy and bicycle (60.0 kg). Substitution into the equation gives $KE = \tfrac{1}{2}(60.0$ kg$)(8.00$ m/s$)^2$. Solving gives 1920 J.

Work is defined as the product of the applied force and the distance the force moves through. The work done on an object changes some form of energy (potential and/or kinetic and/or internal) of the object and is equal to the total change in energy of the object.

$$W = Fd = \Delta E_T$$

where: $W = work$
$F = force$
$d = displacement\ or\ distance$
$\Delta E_T = change\ in\ total\ energy$

Example: A horizontal force of 5.0 newtons acts on a 3.0-kilogram mass over a distance of 6.0 meters along a horizontal, frictionless surface. What is the change in kinetic energy of the mass during its movement over the 6.0-meter distance?

(1) 6.0 J   (2) 15 J   (3) 30. J   (4) 90. J

Solution: 3 The change in kinetic energy of the object is equal to the work done on the object in pushing it along the horizontal surface. There is no change in the potential energy of the object since the surface is horizontal. Substituting into the equation of work gives $W = (5.0\ N)(6.0\ m) = 30.\ J$.

Power is defined as the rate at which work is done or energy is transferred and is calculated by dividing the amount of work done by the time in which that work is done. The third term in this equation is obtained by substituting Fd for W. The last term is obtained from the definition of average speed as distance divided by time.

$$P = \frac{W}{t} = \frac{Fd}{t} = F\overline{v}$$

where: $P = power$
$W = work$
$t = time\ interval$
$F = force$
$d = displacement\ or\ distance$
$\overline{v} = average\ velocity\ or\ average\ speed$

Example: A 40.-kilogram student runs up a staircase to a floor that is 5.0 meters higher than her starting point in 7.0 seconds. The student's power output is

(1) 29 W   (2) 280 W   (3) 1.4×10^3 W   (4) 1.4×10^4 W

Solution: 2 Under Mechanics, find the equations $g = F_g/m$ and $P = Fd/t$. In running up a staircase, the student must use a force equal to her weight. Solving the first equation for F_g gives $F_g = mg$. Substituting this for F in the power equation gives $P = mgd/t$ where d is the vertical height. Substitution gives $P = (40.\ kg)(9.81\ m/s^2)(5.0\ m)/(7.0\ s)$. Solving yields 280 W.

The total energy of an object is the sum of the objects potential energy, kinetic energy and internal energy. Internal energy refers to the heat content of the object. It is changed when work is done against friction.

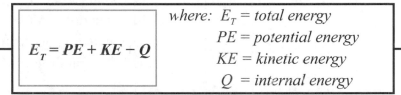

$$E_T = PE + KE - Q$$

where: E_T = total energy
PE = potential energy
KE = kinetic energy
Q = internal energy

Example: A 3.0-kilogram object is placed on a frictionless track at point A and released from rest. (Assume the gravitational potential energy of the system to be zero at point C.)

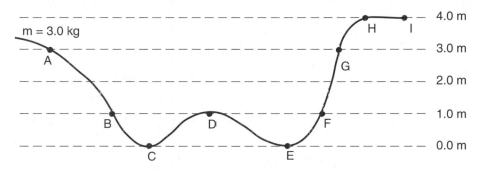

Calculate the kinetic energy of the object at point B. [Show all work, including the equation and substitution with units.]

Solution:

$$E_T = PE + KE + Q$$
$$KE = mg\Delta h$$
$$KE = (3.0 \text{ kg})(9.81 \text{ m/s}^2)(3.0 \text{ m} - 1.0 \text{ m})$$
$$KE = 59 \text{ J} \quad or \quad 58.9 \text{ J}$$

Explanation: Since the system is frictionless, $Q = 0$. The total energy of the system must remain the same. Therefore, the decrease in PE between points A and B must equal the increase in KE between points A and B ($KE = \Delta PE = mg\Delta h$). The $KE = 0$ at point A since the object is at rest at point A. The change in height as the object moves from point A to point B is 2.0 m. Substitute into the equation and calculate KE.

Work, Energy and Power – Additional Information:

• Work and energy are equivalent quantities.

• When there is no change in internal energy or heat ($Q = 0$), the change in potential energy is equal to the change in kinetic energy. As one increases, the other decreases, and vice versa, such that the total energy remains constant (energy is conserved).

1. Which combination of fundamental units can be used to express energy?

   (1) kg•m/s   (3) kg•m/s²
   (2) kg•m²/s   (4) kg•m²/s²   1 _____

Note: Question 2 has only three choices.

2. As an object falls freely, the kinetic energy of the object

   (1) decreases
   (2) increases
   (3) remains the same   2 _____

3. A car travels at constant speed v up a hill from point A to point B, as shown in the diagram below. As the car travels from A to B, its gravitational potential energy

   ```
              B
           v
      ⬤━━━━
     A
   ─────────────────
        Horizontal
   ```

   (1) increases and its kinetic energy decreases
   (2) increases and its kinetic energy remains the same
   (3) remains the same and its kinetic energy decreases
   (4) remains the same and its kinetic energy remains the same   3 _____

4. A 0.50-kilogram ball is thrown vertically upward with an initial kinetic energy of 25 joules. Approximately how high will the ball rise? [Neglect air resistance.]

   (1) 2.6 m   (3) 13 m
   (2) 5.1 m   (4) 25 m   4 _____

5. A 2.0-kilogram block sliding down a ramp from a height of 3.0 meters above the ground reaches the ground with a kinetic energy of 50. joules. The total work done by friction on the block as it slides down the ramp is approximately

   (1) 6 J   (3) 18 J
   (2) 9 J   (4) 44 J   5 _____

6. What is the maximum height to which a 1200-watt motor could lift an object weighing 200. newtons in 4.0 seconds?

   (1) 0.67 m   (3) 6.0 m
   (2) 1.5 m   (4) 24 m   6 _____

7. As shown in the diagram below, a student exerts an average force of 600. newtons on a rope to lift a 50.0-kilogram crate a vertical distance of 3.00 meters. Compared to the work done by the student, the gravitational potential energy gained by the crate is

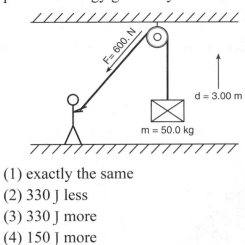

   (1) exactly the same
   (2) 330 J less
   (3) 330 J more
   (4) 150 J more   7 _____

8. A box is pushed to the right with a varying horizontal force. The accompanying graph represents the relationship between the applied force and the distance the box moves. What is the total work done in moving the box 6.0 meters?

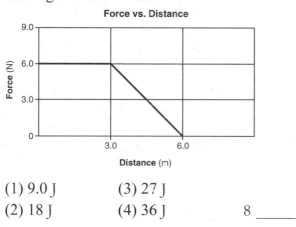

Force vs. Distance

(1) 9.0 J   (3) 27 J

(2) 18 J   (4) 36 J   8 _____

9. A 70.-kilogram cyclist develops 210 watts of power while pedaling at a constant velocity of 7.0 meters per second east. What average force is exerted eastward on the bicycle to maintain this constant speed?

(1) 490 N   (3) 3.0 N

(2) 30. N   (4) 0 N   9 _____

10. Which graph best represents the relationship between the power required to raise an elevator and the speed at which the elevator rises?

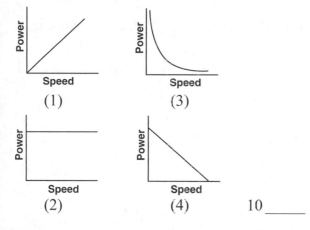

(1)

(3)

(2)

(4)   10 _____

11. The table below lists the mass and speed of each of four objects.

Data Table

Objects	Mass (kg)	Speed (m/s)
A	1.0	4.0
B	2.0	2.0
C	0.5	4.0
D	4.0	1.0

Which two objects have the same kinetic energy?

(1) A and D   (3) A and C

(2) B and D   (4) B and C   11 _____

12. An object moving at a constant speed of 25 meters per second possesses 450 joules of kinetic energy. What is the object's mass?

(1) 0.72 kg   (3) 18 kg

(2) 1.4 kg   (4) 36 kg   12 _____

13. The diagram below shows a moving, 5.00-kilogram cart at the foot of a hill 10.0 meters high. For the cart to reach the top of the hill, what is the minimum kinetic energy of the cart in the position shown? [Neglect energy loss due to friction.]

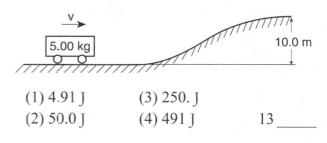

(1) 4.91 J   (3) 250. J

(2) 50.0 J   (4) 491 J   13 _____

14. A constant force of 1900 newtons is required to keep an automobile having a mass of 1.0×10^3 kilograms moving at a constant speed of 20. meters per second. The work done in moving the automobile a distance of 2.0×10^3 meters is

(1) 2.0×10^4 J   (3) 2.0×10^6 J
(2) 3.8×10^4 J   (4) 3.8×10^6 J   14 ____

15. A 1.0-kilogram rubber ball traveling east at 4.0 meters per second hits a wall and bounces back toward the west at 2.0 meters per second. Compared to the kinetic energy of the ball before it hits the wall, the kinetic energy of the ball after it bounces off the wall is

(1) one-fourth as great
(2) one-half as great
(3) the same
(4) four times as great   15 ____

Base your answers to questions 16 a and b on the information and accompanying diagram.

A 250.-kilogram car is initially at rest at point A on a roller coaster track. The car carries a 75-kilogram passenger and is 20. meters above the ground at point A. [Neglect friction.]

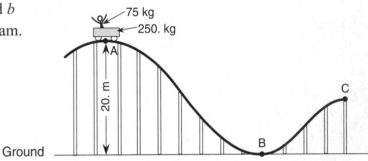

16. a) Calculate the total gravitational potential energy, relative to the ground, of the car and the passenger at point A. [Show all work, including the equation and substitution with units.]

b) Calculate the speed of the car and passenger at point B. [Show all work, including the equation and substitution with units.]

Base your answers to questions 17 *a* and *b* on the information below.

A 65-kilogram pole vaulter wishes to vault to a height of 5.5 meters.

17. *a)* Calculate the *minimum* amount of kinetic energy the vaulter needs to reach this height if air friction is neglected and all the vaulting energy is derived from kinetic energy. [Show all work, including the equation and substitution with units.]

b) Calculate the speed the vaulter must attain to have the necessary kinetic energy. [Show all work, including the equation and substitution with units.]

18. A book sliding across a horizontal tabletop slows until it comes to rest. Describe what change, if any, occurs in the book's kinetic energy and internal energy as it slows.

19. Which quantity is a measure of the rate at which work is done?

    (1) energy   (3) momentum
    (2) power   (4) velocity   19 _____

20. A 1.0-kilogram book resting on the ground is moved 1.0 meter at various angles relative to the horizontal. In which direction does the 1.0-meter displacement produce the greatest increase in the book's gravitational potential energy?

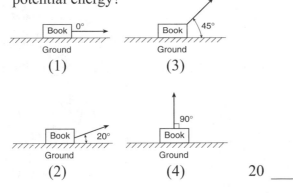

21. During an emergency stop, a 1.5×10^3-kilogram car lost a total of 3.0×10^5 joules of kinetic energy. What was the speed of the car at the moment the brakes were applied?

    (1) 10. m/s   (3) 20. m/s
    (2) 14 m/s   (4) 25 m/s   21 _____

22. In raising an object vertically at a constant speed of 2.0 meters per second, 10. watts of power is developed. The weight of the object is

    (1) 5.0 N   (3) 40. N
    (2) 20. N   (4) 50. N   22 _____

Note: Question 23 has only three choices.

23. Two weightlifters, one 1.5 meters tall and one 2.0 meters tall, raise identical 50.-kilogram masses above their heads. Compared to the work done by the weightlifter who is 1.5 meters tall, the work done by the weightlifter who is 2.0 meters tall is

    (1) less   (2) greater   (3) the same   23 _____

24. The graph below shows the relationship between the work done by a student and the time of ascent as the student runs up a flight of stairs.

    The slope of the graph would have units of

    (1) joules   (3) watts
    (2) seconds   (4) newtons   24 _____

25. A 95-kilogram student climbs 4.0 meters up a rope in 3.0 seconds. What is the power output of the student?

    (1) 1.3×10^2 W
    (2) 3.8×10^2 W
    (3) 1.2×10^3 W
    (4) 3.7×10^3 W   25 _____

26. What is the average power developed by a motor as it lifts a 400.-kilogram mass at constant speed through a vertical distance of 10.0 meters in 8.0 seconds?

    (1) 320 W   (3) 4,900 W
    (2) 500 W   (4) 32,000 W   26 _____

Base your answers to questions 27 a, b, and c on the information and accompanying table. The table lists the kinetic energy of a 4.0-kilogram mass as it travels in a straight line for 12.0 seconds.

Time (seconds)	Kinetic Energy (joules)
0.0	0.0
2.0	8.0
4.0	18
6.0	32
10.0	32
12.0	32

Directions: Using the information in the data table, construct a graph on the accompanying grid, following the directions below.

27. *a*) Mark an appropriate scale on the axis labeled "Kinetic Energy (J)."

b) Plot the data points for kinetic energy versus time.

c) Calculate the speed of the mass at 10.0 seconds. [Show all work, including the equation and substitution with units.]

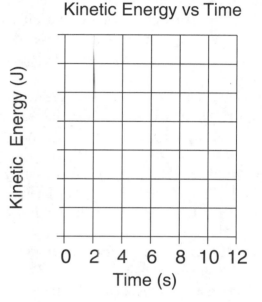

Kinetic Energy vs Time

Base your answers to questions 28 *a* and *b* on the information below.

A 75-kilogram athlete jogs 1.8 kilometers along a straight road in 1.2×10^3 seconds.

28. *a*) Determine the average speed of the athlete in meters per second.

b) Calculate the average kinetic energy of the athlete. [Show all work, including the equation and substitution with units.]

Answers

Set 1

1. **4**   Energy, and its equivalent, work, are measured in joules (J). In the reference table, find the equation $KE = \frac{1}{2}mv^2$. The unit of kinetic energy is the unit of mass (kg) multiplied by the square of the unit of speed (m/s). In fundamental units, energy is expressed in $kg \cdot m^2/s^2$.

2. **2**   As an object falls freely, mechanical energy (the total potential and kinetic energy of an object) remains constant. As the object falls, it's gravitational potential energy decreases due to a decrease in the height above the Earth's surface. As the gravitational potential energy decreases, it's kinetic energy must increase.

3. **2**   Under Mechanics, find the equations $\Delta PE = mg\Delta h$ and $KE = \frac{1}{2}mv^2$. As the car travels up the hill, Δh increases, therefore increasing the potential energy of the car. Since the car is moving at a constant speed, the kinetic energy of the car remains the same.

4. **2**   As the ball rises, its kinetic energy is converted to potential energy. At the highest point, all of the initial kinetic energy has been converted to potential energy. Under Mechanics, find the equation $\Delta PE = mg\Delta h$. Substitution gives 25 J = (0.50 kg)(9.81 m/s²)Δh. Solving gives Δh = 5.1 m.

5. **2**   Ideally, the mechanical energy of an object (the sum of the potential and kinetic energy of an object) remains constant. The decrease in potential energy of the object as it slides down the ramp should equal the increase in kinetic energy of the object. Any difference will be due to work done against friction during the slide down the ramp. Under Mechanics, find the equation $\Delta PE = mg\Delta h$. The value of g is found on the List of Physical Constants in the reference table. Substituting into the equation gives ΔPE = (2.0 kg)(9.81 m/s²)(3.0 m) = 58.8 J. The increase in kinetic energy of the object should be 58.8 J. The difference (58.8 J – 50.0 J = 8.8 J) is the work done against friction.

6. **4**   Under Mechanics, find the equation $P = \dfrac{Fd}{t}$. The force needed to lift the object is equal to the weight of the object. Substituting into the equation gives $1200\text{ W} = \dfrac{(200\text{ N})(d)}{4.0\text{ s}}$ Solving for d gives an answer of 24 m.

7. **2**   Under Mechanics, find the equation $W = Fd$. The student exerts a force of 600. N through a distance of 3.00 m in lifting the mass. The work done by the student is therefore W = (600. N)(3.0 m) = 1800 J. Also find the equation $\Delta PE = mg\Delta h$. The change in potential energy of the mass as it is lifted is ΔPE = (50.0 kg)(9.81 m/s²)(3.00 m) = 1470 J. Comparing the values of the potential energy and the work done shows that the potential energy is 330 J less (1800 J – 1470 J).

8. 3   The total work done is the sum of the work done between 0 m and 3.0 m and the work done between 3.0 m and 6.0 m. In the reference table, find the equation $W = Fd$. From 0 m to 3.0 m, the force is constant so $W = (6.0 \text{ N}) (3.0 \text{ m}) = 18$ J. From 3.0 m to 6.0 m, the force changes from 6.0 N to 0 N. To calculate the work done, use the average force, which is $\frac{(6.0 \text{ N} + 0 \text{ N})}{2} = 3.0$ N. The work done is $W = (3.0 \text{ N})(3.0 \text{ m}) = 9.0$ J. The total work done is 18 J $+ 9.0$ J $= 27$ J.

9. 2   Under Mechanics, find the equation $P = F\bar{v}$. Substitution gives 210 W $= (F)(7.0$ m/s$)$. Solving, F = 30. N.

10. 1   Locate the equation $P = F\bar{v}$. This shows that there is a direct relationship between the power required and the speed at which the elevator is raised. This is indicated by graph 1.

11. 4   Under Mechanics, find the equation $KE = \frac{1}{2} mv^2$. Substituting the mass and speeds of objects A, B, C and D into this equation and solving gives kinetic energies of 8.0 J, 4.0 J, 4.0 J and 2.0 J. Objects B and C have the same kinetic energies.

12. 2   In the reference table, find the equation $KE = \frac{1}{2} mv^2$. Substituting into the equation gives 450 J $= \frac{1}{2}(m)(25$ m/s$)^2$ and m = 1.4 kg.

13. 4   Mechanical energy is conserved in this situation. As the cart moves to the top of the hill, the kinetic energy it has in the position shown is converted to potential energy at the top of the hill. Under Mechanics, find the equation $\Delta PE = mg\Delta h$. Solving for the potential energy at the top gives $\Delta PE = (5.00$ kg$)(9.81$ m/s$^2)(10.0$ m$) = 491$ J. This must then be the kinetic energy of the cart at the bottom of the hill.

14. 4   Find the equation $W = Fd$ under Mechanics. Substitution into the equation yields $W = (1900$ N$)(2.0 \times 10^3$ m$) = 3.8 \times 10^6$ J. The mass and speed are not a factor in this calculation.

15. 1   The direction the ball is traveling does not affect the kinetic energy of the ball. Under Mechanics, find the equation $KE = \frac{1}{2} mv^2$. This shows that the kinetic energy varies directly as the square of the speed of the object. If the speed is halved, the kinetic energy is then one-fourth as great. Using the kinetic energy equation, you can also calculate the kinetic energy of the ball before and after it hits the wall. Substitution into the equation gives $KE = \frac{1}{2} (1.0$ kg$)(4.0$ m/s$)^2$ before and $KE = \frac{1}{2} (1.0$ kg$)(2.0$ m/s$)^2$ after. The KE before is 8.0 J and after the KE is 2.0 J. 2.0 J is one-fourth of 8.0 J.

16. a) $\Delta PE = mg\Delta h$
$\Delta PE = (250.\ \text{kg} + 75\ \text{kg})(9.81\ \text{m/s}^2)(20.\ \text{m})$
$\Delta PE = 6.4 \times 10^4\ \text{J}$

Explanation: Locate the equation $\Delta PE = mg\Delta h$ The mass is the sum of the masses of the car and passenger. Substitution into the equation and solving gives the potential energy at point A.

16. b) $\Delta PE = KE = \frac{1}{2}mv^2$

$v = \sqrt{\dfrac{2\Delta PE}{m}}$

$v = \sqrt{\dfrac{2(6.4 \times 10^4\ \text{J})}{325\ \text{kg}}}$

$v = 20.\ \text{m/s}$

or

$\Delta PE = KE = \frac{1}{2}mv^2$

$6.4 \times 10^4\ \text{J} = \frac{1}{2}(250.\ \text{kg} + 75\ \text{kg})v^2$

$v^2 = 394\ \text{m}^2/\text{s}^2$

$v = 20.\ \text{m/s}$

Explanation: As the car and passenger travel from point A to point B, the potential energy at point A is converted to kinetic energy at point B. The equation for kinetic energy (KE) is found in the reference table under Mechanics. Use the value calculated in question 16a for the kinetic energy at point B.

17. a) $KE = \Delta PE = mg\Delta h$
$KE = (65\ \text{kg})(9.81\ \text{m/s}^2)(5.5\ \text{m})$
$KE = 3.5 \times 10^3\ \text{J}$

Explanation: In the absence of friction, the change in kinetic energy is equal to the change in potential energy. The minimum kinetic energy needed must be equal to the potential energy of the vaulter at the highest point. Under Mechanics, find the equations $KE = \frac{1}{2}\,mv^2$ and $\Delta PE = mg\Delta h$. Combining these equations, $KE = \Delta PE = mg\Delta h$. The value of g is found on the List of Physical Constants.

17. b) $KE = \dfrac{1}{2}mv^2$

$v = \sqrt{\dfrac{2KE}{m}}$

$v = \sqrt{\dfrac{2(3.5 \times 10^3\ \text{J})}{65\ \text{kg}}}$

$v = 10.\ \text{m/s}$

Explanation: Find the equation $KE = \frac{1}{2}\,mv^2$. Using the value for kinetic energy calculated in question 17a, 3500 J = $\frac{1}{2}$ (65 kg)(v)^2. Solving, $v = 10.\ \text{m/s}$.

18. Acceptable response: The kinetic energy decreases and the internal energy increases.

Explanation: As the book slows down to rest, the kinetic energy of the book decreases due to the decrease in speed ($KE = \frac{1}{2}\,mv^2$). The kinetic energy is changed into internal energy, causing it to increase. There is no change in potential energy since the book is moving along a horizontal surface.

Circular Motion: Uniform circular motion is the motion of an object along a path of constant radius at a constant speed.

Circular motion is caused by what is called a centripetal force. This is a force that acts perpendicular to the direction of motion of an object. It is equal to the product of the mass of an object and its centripetal acceleration. The acceleration of an object in circular motion is due to the change in direction of the objects velocity. This equation is an application of Newton's Second Law to circular motion.

$$F_c = ma_c$$

where: F_c = centripetal force
m = mass
a_c = centripetal acceleration

Example: A 20. kg object has a centripetal acceleration of 2.5 m/s². What is the centripetal force acting on the object?

(1) 8.0 N   (2) 0.125 N   (3) 50. N   (4) 75. N

Solution: 3 The mass and centripetal acceleration of the object are given. Substitution directly into the equation gives F_c = (20. kg)(2.5 m/s²) = 50. N

The acceleration produced by a centripetal force is called centripetal acceleration and is calculated by dividing the square of the speed of the object by the radius of the circular path.

$$a_c = v^2/r$$

where: a_c = centripetal acceleration
v = velocity or speed
r = radius or distance between centers

Example: A 2.0×10^3 -kilogram car travels at a constant speed of 12 meters per second around a circular curve of radius 30. meters. What is the magnitude of the centripetal acceleration of the car as it goes around the curve?

(1) 0.40 m/s²   (2) 4.8 m/s²   (3) 800 m/s²   (4) 9,600 m/s²

Solution: 2 The velocity of the car and the radius of the curve are given. Substitution yields $a_c = \dfrac{(12 \text{ m/s})^2}{(30. \text{ m})}$ and solving gives a_c = 4.8 m/s².

Circular Motion – Additional Information:

• The centripetal force and centripetal acceleration are directed toward the center of the circular path.

1. The diagram below shows a student seated on a rotating circular platform, holding a 2.0-kilogram block with a spring scale. The block is 1.2 meters from the center of the platform. The block has a constant speed of 8.0 meters per second. [Frictional forces on the block are negligible.]

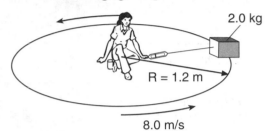

2.0 kg

R = 1.2 m

8.0 m/s

Which statement best describes the block's movement as the platform rotates?

(1) Its velocity is directed tangent to the circular path, with an inward acceleration.
(2) Its velocity is directed tangent to the circular path, with an outward acceleration.
(3) Its velocity is directed perpendicular to the circular path, with an inward acceleration.
(4) Its velocity is directed perpendicular to the circular path, with an outward acceleration.   1 _____

2. A child is riding on a merry-go-round. As the speed of the merry-go-round is doubled, the magnitude of the centripetal force acting on the child

(1) remains the same
(2) is doubled
(3) is halved
(4) is quadrupled   2 _____

3. A 0.50-kilogram object moves in a horizontal circular path with a radius of 0.25 meter at a constant speed of 4.0 meters per second. What is the magnitude of the object's acceleration?

(1) 8.0 m/s^2 (3) 32 m/s^2
(2) 16 m/s^2 (4) 64 m/s^2 3 _____

Base your answers to questions 4 and 5 on the information and the diagram below.

The diagram shows the top view of a 65-kilogram student at point A on an amusement park ride. The ride spins the student in a horizontal circle of radius 2.5 meters, at a constant speed of 8.6 meters per second. The floor is lowered and the student remains against the wall without falling to the floor.

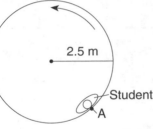

2.5 m

Student
A

4. Which vector best represents the direction of the centripetal acceleration of the student at point A?

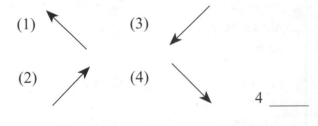

(1)   (3)

(2)   (4)   4 _____

5. The magnitude of the centripetal force acting on the student at point A is approximately

(1) 1.2×10^4 N   (3) 2.2×10^2 N
(2) 1.9×10^3 N   (4) 3.0×10^1 N   5 _____

6. In the accomapanying diagram, S is a point on a car tire rotating at a constant rate.

Which graph best represents the magnitude of the centripetal acceleration of point S as a function of time?

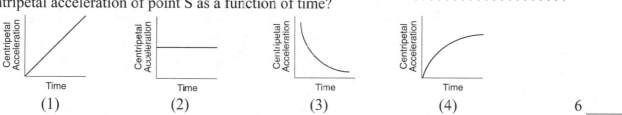

(1) (2) (3) (4) 6 _____

7. Calculate the magnitude of the centripetal force acting on Earth as it orbits the Sun, assuming a circular orbit and an orbital speed of 3.00×10^4 meters per second. [Show all work, including the equation and substitution with units.]

Base your answers to questions 8 *a, b, c* on the information below.

The combined mass of a race car and its driver is 600. kilograms. Traveling at constant speed, the car completes one lap around a circular track of radius 160 meters in 36 seconds.

8. *a*) Calculate the speed of the car. [Show all work, including the equation and substitution with units.]

b) On the accompanying diagram, draw an arrow to represent the direction of the net force acting on the car when it is in position *A*.

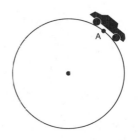

c) Calculate the magnitude of the centripetal acceleration of the car. [Show all work, including the equation and substitution with units.]

9. Centripetal force F_c acts on a car going around a curve. If the speed of the car were twice as great, the magnitude of the centripetal force necessary to keep the car moving in the same path would be

   (1) F_c (3) $\frac{F_c}{2}$

   (2) $2F_c$ (4) $4F_c$ 9 _____

10. The diagram below represents a 0.40-kilogram stone attached to a string. The stone is moving at a constant speed of 4.0 meters per second in a horizontal circle having a radius of 0.80 meter.

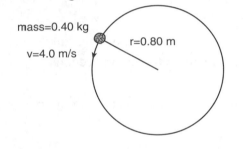

mass=0.40 kg
v=4.0 m/s
r=0.80 m

    The magnitude of the centripetal acceleration of the stone is

    (1) 0.0 m/s²   (3) 5.0 m/s²
    (2) 2.0 m/s²   (4) 20. m/s²   10 _____

11. A 1750-kilogram car travels at a constant speed of 15.0 meters per second around a horizontal, circular track with a radius of 45.0 meters. The magnitude of the centripetal force acting on the car is

    (1) 5.00 N   (3) 8750 N
    (2) 583 N   (4) 3.94 × 10⁵ N   11 _____

12. A 1.0×10^3-kilogram car travels at a constant speed of 20. meters per second around a horizontal circular track. Which diagram correctly represents the direction of the car's velocity (v) and the direction of the centripetal force (F_c) acting on the car at one particular moment?

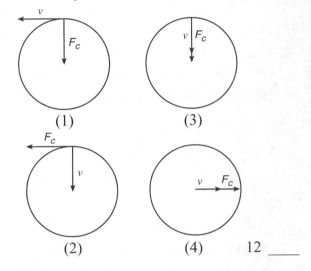

(1)   (3)

(2)   (4)   12 _____

13. Which graph best represents the relationship between the magnitude of the centripetal acceleration and the speed of an object moving in a circle of constant radius?

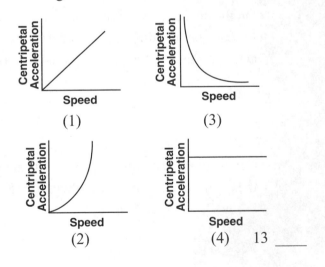

(1)   (3)

(2)   (4)   13 _____

Base your answers to questions 14 *a* and *b* on the information and accompanying diagram.

A 1200-kilogram car traveling at a constant speed of 9.0 meters per second turns at an intersection. The car follows a horizontal circular path with a radius of 25 meters to point *P*.

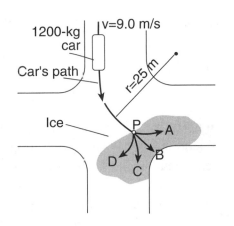

14. *a*) What is the magnitude of the centripetal force acting on the car as it travels around the circular path?

_____

b) At point *P*, the car hits an area of ice and loses all frictional force on its tires. Which path does the car follow on the ice?

_____

Base your answers to questions 15 *a*, *b* and *c* on the information below and accompanying diagram.

A 1.50-kilogram cart travels in a horizontal circle of radius 2.40 meters at a constant speed of 4.00 meters per second.

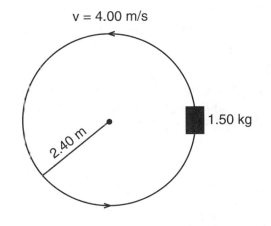

15. *a*) Calculate the time required for the cart to make one complete revolution. [Show all work, including the equation and substitution with units.]

b) Describe a change that would quadruple the magnitude of the centripetal force.

_____

_____

c) On the diagram above, draw an arrow to represent the direction of the acceleration of the cart in the position shown. Label the arrow *a*.

1. 1   In uniform circular motion, the velocity is always tangent to the circular path and the object undergoes an acceleration directed toward the center of the circular path (called centripetal acceleration).

2. 4   Under Mechanics, find the equations $a_c = v^2/r$ and $F_c = ma_c$. Combining these equations gives $F_c = mv^2/r$. This equation indicates that the centripetal force varies directly with the square of the speed of the object. Therefore, if the speed is doubled, the centripetal force quadruples.

3. 4   Under Mechanics, find the equation $a_c = v^2/r$. Substitution gives $a_c = (4.0 \text{ m/s})^2/(0.25 \text{ m})$. Solving, $a_c = 64 \text{ m/s}^2$.

4. 1   The word centripetal means pointing toward the center or center seeking. Therefore the centripetal acceleration must point toward the center of the circular path.

5. 2   Under Mechanics, find the equations $F_c = ma_c$ and $a_c = v^2/r$. Substituting for a_c in the first equation gives $F_c = mv^2/r$. Substitution into this equation gives $F_c = (65 \text{ kg})(8.6 \text{ m/s})^2/(2.5 \text{ m})$. Solving, $F_c = 1.9 \times 10^3 \text{ N}$.

6. 2   Using the reference table, locate the equation $a_c = v^2/r$. Since the speed of point S is constant and the radius of the path point S follows is constant, the centripetal acceleration of point S is constant with time.

7.

$$F_c = ma_c \quad \text{and} \quad a_c = \frac{v^2}{r}$$

$$F_c = \frac{mv^2}{r}$$

$$F_c = \frac{\left(5.98 \times 10^{24} \text{ kg}\right)\left(3.00 \times 10^4 \text{ m/s}\right)^2}{1.5 \times 10^{11} \text{ m}}$$

$$F_c = 3.59 \times 10^{22} \text{ N}$$

Explanation: For this problem one must use the equations $F_c = ma_c$ and $a_c = v^2/r$. Substituting for a_c in the first equation gives $F_c = mv^2/r$. The mass of the Earth and the mean distance – Earth to the Sun is found in the List of Physical Constants.

8. *a)*

$$\bar{v} = \frac{d}{t}$$

$$\bar{v} = \frac{2\pi r}{t}$$

$$\bar{v} = \frac{2\pi(160 \text{ m})}{36 \text{ s}}$$

$$\bar{v} = 28 \text{ m/s} \quad or \quad 27.9 \text{ m/s}$$

Explanation: Under Mechanics, find the equation $\bar{v} = d/t$. In this equation, d is the circumference of the circular track. Under Geometry and Trigonometry, find the equation $C = 2\pi r$. The first equation then may be written $\bar{v} = 2\pi r/t$. Substitute into the equation and solve for $\bar{v}$. Since the speed is constant, the average speed is the speed of the race car.

8. *b)*

Explanation: The net force acting on the race car to produce the circular motion is the centripetal force. By definition, the centripetal force acts toward the center of the circle.

8. *c)*

$$a_c = \frac{v^2}{r}$$

$$a_c = \frac{(28 \text{ m/s})^2}{160 \text{ m}}$$

$$a_c = 4.9 \text{ m/s}^2$$

Explanation: Under Mechanics, find the equation $a_c = v^2/r$. Substitute into the equation using the speed calculated in question 8*a* and solve for a_c.

Other Equations: These two equations are used to calculate components of a given force.

The y or vertical component of a given vector A at an angle θ with the horizontal is the product of the magnitude of that vector and the sine of the angle θ.

$$A_y = A \sin \theta$$

where: A = any vector quantity
θ = angle

Example: A football player kicks a ball with an initial velocity of 25 meters per second at an angle of 53° above the horizontal. The vertical component of the initial velocity of the ball is

(1) 25 m/s   (2) 20. m/s   (3) 15 m/s   (4) 10. m/s

Solution: 2 Substitution into the equation gives A_y = (25 m/s)(sin 53°) = (25 m/s)(0.799). Solving, A_y = 19.9 m/s.

The x or horizontal component of a given vector A at an angle θ with the horizontal is the product of the magnitude of that vector and the cosine of the angle θ.

$$A_x = A \cos \theta$$

where: A = any vector quantity
θ = angle

Example: A golf ball is propelled with an initial velocity of 60. meters per second at 37° above the horizontal. The horizontal component of the golf ball's initial velocity is

(1) 30. m/s   (2) 36 m/s   (3) 40. m/s   (4) 48 m/s

Solution: 4 A_x represents the horizontal component of the initial velocity of the golf ball. Substituting into the equation gives A_x = (60. m/s)(cos 37°). Solving, A_x = 48 m/s.

Other Equations – Additional Information:

• Components of a given vector are two or more vectors that can replace or have the same effect as the original vector.

1. An airplane flies with a velocity of 750. kilometers per hour, 30.0° south of east. What is the magnitude of the eastward component of the plane's velocity?

    (1) 866 km/h   (3) 433 km/h

    (2) 650. km/h   (4) 375 km/h   1 _____

2. A soccer player kicks a ball with an initial velocity of 10. meters per second at an angle of 30.° above the horizontal. The magnitude of the horizontal component of the ball's initial velocity is

    (1) 5.0 m/s   (3) 9.8 m/s

    (2) 8.7 m/s   (4) 10. m/s   2 _____

3. A projectile is fired with an initial velocity of 120. meters per second at an angle, θ, above the horizontal. If the projectile's initial horizontal speed is 55 meters per second, the angle measures approximately

    (1) 13°   (2) 27°   (3) 63°   (4) 75°   3 _____

4. An outfielder throws a baseball to the first baseman at a speed of 19.6 meters per second and an angle of 30.° above the horizontal. Which pair represents the initial horizontal velocity (v_x) and initial vertical velocity (v_y) of the baseball?

    (1) $v_x = 17.0$ m/s, $v_y = 9.80$ m/s

    (2) $v_x = 9.80$ m/s, $v_y = 17.0$ m/s

    (3) $v_x = 19.4$ m/s, $v_y = 5.90$ m/s

    (4) $v_x = 19.6$ m/s, $v_y = 19.6$ m/s   4 _____

Base your answers to questions 5a and b on the information below.

A kicked soccer ball has an initial velocity of 25 meters per second at an angle of 40.° above the horizontal, level ground. [Neglect friction.]

5. a) Calculate the magnitude of the vertical component of the ball's initial velocity. [Show all work, including the equation and substitution with units.]

    b) Calculate the maximum height the ball reaches above its initial position. [Show all work, including the equation and substitution with units.]

6. A child kicks a ball with an initial velocity of 8.5 meters per second at an angle of 35° with the horizontal, as shown. The ball has an initial vertical velocity of 4.9 meters per second and a total time of flight of 1.0 second. [Neglect air resistance.]

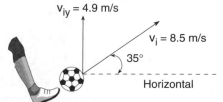

The horizontal component of the ball's initial velocity is approximately

(1) 3.6 m/s   (3) 7.0 m/s
(2) 4.9 m/s   (4) 13 m/s   6 _____

7. A golf ball is hit with an initial velocity of 15 meters per second at an angle of 35 degrees above the horizontal. What is the vertical component of the golf ball's initial velocity?

(1) 8.6 m/s   (3) 12 m/s
(2) 9.8 m/s   (4) 15 m/s   7 _____

8. A catapult launches a rock with an initial speed of 80. m/s at an angle θ above the horizontal. If the rock's initial horizontal speed is 25 m/s, then angle θ measures approximately

(1) 26°   (3) 56°
(2) 34°   (4) 72°   8 _____

9. A golf ball leaves a golf club with an initial velocity of 40.0 meters per second at an angle of 40.° with the horizontal.

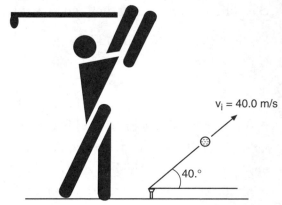

What is the vertical component of the golf ball's initial velocity?

(1) 25.7 m/s   (3) 40.0 m/s
(2) 30.6 m/s   (4) 61.3 m/s   9 _____

10. The vector diagram below represents the horizontal component, F_H, and the vertical component, F_V, of a 24-newton force acting at 35° above the horizontal.

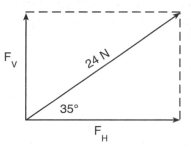

What are the magnitudes of the horizontal and vertical components?

(1) F_H = 3.5 N and F_V = 4.9 N
(2) F_H = 4.9 N and F_V = 3.5 N
(3) F_H = 14 N and F_V = 20. N
(4) F_H = 20. N and F_V = 14 N   10 _____

Base your answers to questions 11a and b on the information and diagram below.

A force of 60. newtons is applied to a rope to pull a sled across a horizontal surface at a constant velocity. The rope is at an angle of 30. degrees above the horizontal.

F = 60. N

30.°

Horizontal surface

11. a) Calculate the magnitude of the component of the 60.-newton force that is parallel to the horizontal surface. [Show all work, including the equation and substitution with units.]

b) Determine the magnitude of the frictional force acting on the sled. _____N

Base your answers to questions 12a, b and c on the information and diagram below.

A projectile is fired from the ground with an initial velocity of 250. meters per second at an angle of 60.° above the horizontal.

P   Horizontal

12. a) On the diagram above, use a protractor and ruler to draw a vector to represent the initial velocity of the projectile. Begin the vector at point P, and use a scale of 1.0 centimeter = 50. meters per second.

b) Determine the horizontal component of the initial velocity.

c) Explain why the projectile has no acceleration in the horizontal direction. [Neglect air friction.]

_____

_____

1. 2   Under Mechanics, find the equation $A_x = A \cos \theta$. Since the plane is flying southeast, the eastward component will be along the x direction. Substituting and solving,
$A_x = (750.\ \text{km/h})(\cos 30.0°) = 650.\ \text{km/h}$.

2. 2   Under Mechanics, find the equation $A_x = A \cos \theta$. Substitution into the equation gives
$A_x = (10.\ \text{m/s})(\cos 30°) = (10.\ \text{m/s})(0.866) = 8.7\ \text{m/s}$.

3. 3   Under Mechanics, find the equation $A_x = A \cos \theta$. Substitution into the equation gives
$55\ \text{m/s} = (120\ \text{m/s})(\cos \theta)$. Solving, $\cos \theta = 0.46$. The angle whose cosine is 0.46 is 63°.

4. 1   Under Mechanics, find the equations $A_x = A \cos \theta$ and $A_y = A \sin \theta$ where A_x and A_y represent the horizontal and vertical components of that vector quantity, in this case a velocity. Substitution into the equations gives $v_x = (19.6\ \text{m/s})(\cos 30°)$ and $v_y = (19.6\ \text{m/s})(\sin 30°)$. Solving yields $v_x = 17.0\ \text{m/s}$ and $v_y = 9.80\ \text{m/s}$.

5. a) Answer: $A_y = 16\ \text{m/s}$

    Explanation: Under Mechanics, find the equation $A_y = A \sin \theta$. A_y represents the vertical component of the initial velocity of the ball. Substituting and solving gives
    $A_y = (25\ \text{m/s})(\sin 40°) = (25\ \text{m/s})(0.64) = 16\ \text{m/s}$.

    b) Answer: $d = 13\ \text{m}$

    Explanation: Under Mechanics, find the equation $v_f^2 = v_i^2 + 2ad$. In the vertical, $v_i = 16\ \text{m/s}$ (your value of the vertical component calculated in question 5a) and $v_f = 0$, the speed at the highest point. The acceleration is that due to gravity (g), found on the List of Physical Constants. Substitution gives $0 = (16\ \text{m/s})^2 + 2(9.81\ \text{m/s}^2)(d)$. Solving, $d = 13\ \text{m}$.

Waves – Wave Speed, Period and Frequency

A **Wave** is a vibratory disturbance that carries energy from a source. If a wave is transmitted through a material medium, such as sound, the particles of the medium vibrate to and fro about an equilibrium or rest position. They do not travel along with the wave. Electromagnetic waves do not require a material medium for transmission.

Wave Speed, Period and Frequency: The speed of a wave is the rate at which the waves travel from its source. Characteristics of a wave are its wavelength, frequency and period.

The wavelength is the distance between two consecutive points on a wave that are in phase.

The period of a wave is defined as the time needed to produce one wave or pulse.

The frequency of a wave is the number of waves produced per unit time.

The speed at which a wave travels from one point to another is the product of the frequency and wavelength of the wave.

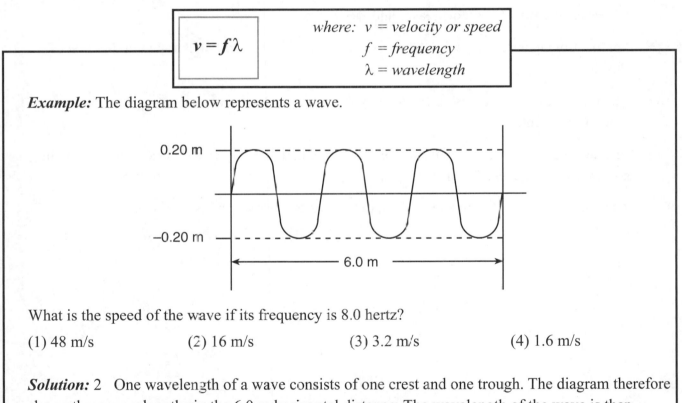

$$v = f\lambda$$

where: v = velocity or speed
f = frequency
λ = wavelength

Example: The diagram below represents a wave.

0.20 m

−0.20 m

6.0 m

What is the speed of the wave if its frequency is 8.0 hertz?

(1) 48 m/s   (2) 16 m/s   (3) 3.2 m/s   (4) 1.6 m/s

Solution: 2   One wavelength of a wave consists of one crest and one trough. The diagram therefore shows three wavelengths in the 6.0 m horizontal distance. The wavelength of the wave is then 2.0 m. Substitution into the equation $v = f\lambda$ gives $v = (8.0 \text{ Hz})(2.0 \text{ m}) = 16$ m/s.

The period and frequency of a wave are reciprocal quantities.

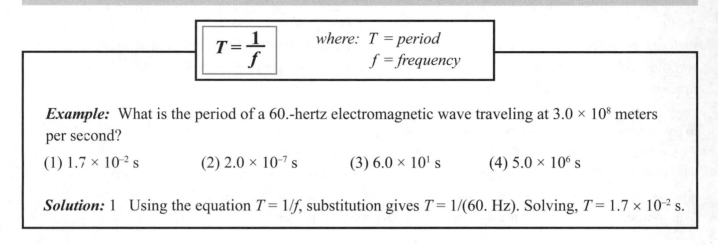

$$T = \frac{1}{f}$$

where: T = period
f = frequency

Example: What is the period of a 60.-hertz electromagnetic wave traveling at 3.0×10^8 meters per second?

(1) 1.7×10^{-2} s   (2) 2.0×10^{-7} s   (3) 6.0×10^{1} s   (4) 5.0×10^6 s

Solution: 1   Using the equation $T = 1/f$, substitution gives $T = 1/(60.\text{ Hz})$. Solving, $T = 1.7 \times 10^{-2}$ s.

Wave Speed, Period and Frequency – Additional Information:

- The speed of a wave depends upon the properties of the medium and is constant for a given wave in a uniform medium. Therefore, as shown in the equation $v = f\lambda$, as the frequency increases, the wavelength decreases and vice versa.

- For electromagnetic waves in a vacuum or air, $v = c$. The value of c is found on the List of Physical Constants.

- The frequency of a wave is determined by the source of the wave, and once generated, remains constant.

- The amplitude of a wave, a measure of the energy carried by the wave, is determined by the source and is independent of the frequency of the wave.

- The unit of frequency is the hertz (Hz), defined as one per second (1/s or s^{-1}).

Set 1 — Wave Speed, Period and Frequency

1. A motor is used to produce 4.0 waves each second in a string. What is the frequency of the waves?

   (1) 0.25 Hz   (3) 25 Hz

   (2) 15 Hz   (4) 4.0 Hz   1 _____

2. In a vacuum, light with a frequency of 5.0×10^{14} hertz has a wavelength of

   (1) 6.0×10^{-21} m

   (2) 6.0×10^{-7} m

   (3) 1.7×10^{6} m

   (4) 1.5×10^{23} m   2 _____

3. A change in the speed of a wave as it enters a new medium produces a change in

   (1) frequency   (3) wavelength

   (2) period   (4) phase   3 _____

Note: question 4 has only three choices.

4. Compared to the period of a wave of red light the period of a wave of green light is

   (1) less   (2) greater   (3) the same   4 _____

5. A surfacing whale in an aquarium produces water wave crests having an amplitude of 1.2 meters every 0.40 second. If the water wave travels at 4.5 meters per second, the wavelength of the wave is

   (1) 1.8 m   (3) 3.0 m

   (2) 2.4 m   (4) 11 m   5 _____

6. What is the period of a wave if 20 crests pass an observer in 4 seconds?

   (1) 80 s   (3) 5 s

   (2) 0.2 s   (4) 4 s   6 _____

7. Which type of wave requires a material medium through which to travel?

   (1) sound   (3) television

   (2) radio   (4) x ray   7 _____

---

8. Calculate the wavelength in a vacuum of a radio wave having a frequency of 2.2×10^{6} hertz. [Show all work, including the equation and substitution with units.]

Base your answers to questions 9a, b and c on the information below.

A student generates a series of transverse waves of varying frequency by shaking one end of a loose spring. All the waves move along the spring at a speed of 6.0 meters per second.

9. *a*) Complete the accompanying data table, by determining the wavelengths for the frequencies given.

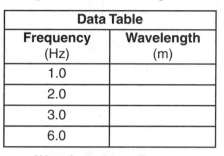

Data Table	
Frequency (Hz)	Wavelength (m)
1.0	
2.0	
3.0	
6.0	

b) On the accompanying grid, plot the data points for wavelength versus frequency.

c) Draw the best-fit line or curve.

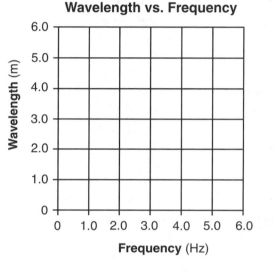

Wavelength vs. Frequency

Base your answers to questions 10a and b on the information and diagram below.

A student standing on a dock observes a piece of wood floating on the water as shown to the right. As a water wave passes, the wood moves up and down, rising to the top of a wave crest every 5.0 seconds.

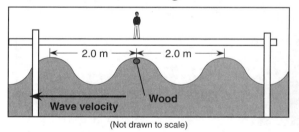

(Not drawn to scale)

10. *a*) Calculate the frequency of the passing water waves. [Show all work, including the equation and substitution with units.]

b) Calculate the speed of the water waves. [Show all work, including the equation and substitution with units.]

Waves

11. A standing wave pattern is produced when a guitar string is plucked. Which characteristic of the standing wave immediately begins to decrease?

    (1) speed   (3) frequency
    (2) wavelength   (4) amplitude   11 ___

Note: Question 12 has only three choices.

12. If the amplitude of a wave is increased, the frequency of the wave will

    (1) decrease
    (2) increase
    (3) remain the same   12 ___

13. Which characteristic is the same for every color of light in a vacuum?

    (1) energy   (3) speed
    (2) frequency   (4) period   13 ___

Note: Question 14 has only three choices.

14. If the amplitude of a wave traveling in a rope is doubled, the speed of the wave in the rope will

    (1) decrease
    (2) increase
    (3) remain the same   14 ___

15. What is the frequency of a wave if its period is 0.25 second?

    (1) 1.0 Hz   (3) 12 Hz
    (2) 0.25 Hz   (4) 4.0 Hz   15 ___

16. Which unit is equivalent to meters per second?

    (1) Hz•s   (3) s/Hz
    (2) Hz•m   (4) m/Hz   16 ___

17. The energy of a water wave is most closely related to its

    (1) frequency   (3) period
    (2) wavelength   (4) amplitude   17 ___

18. If the frequency of a periodic wave is doubled, the period of the wave will be

    (1) halved   (3) quartered
    (2) doubled   (4) quadrupled   18 ___

19. A 512-hertz sound wave travels 100. meters to an observer through air at STP. What is the wavelength of this sound wave?

    (1) 0.195 m   (3) 1.55 m
    (2) 0.646 m   (4) 5.12 m   19 ___

20. The diagram below shows two points, A and B, on a wave train.

    How many wavelengths separate point A and point B?

    (1) 1.0   (3) 3.0
    (2) 1.5   (4) 0.75   20 ___

21. A periodic wave is produced by a vibrating tuning fork. The amplitude of the wave would be greater if the tuning fork were

    (1) struck more softly
    (2) struck harder
    (3) replaced by a lower frequency tuning fork
    (4) replaced by a higher frequency tuning fork   21 ___

22. A student plucks a guitar string and the vibrations produce a sound wave with a frequency of 650 hertz. Calculate the wavelength of the sound wave in air at STP. [Show all work, including the equation and substitution with units.]

Base your answers to questions 23a, b and c on the information below.

A periodic wave traveling in a uniform medium has a wavelength of 0.080 meter, an amplitude of 0.040 meter, and a frequency of 5.0 hertz.

23. a) Determine the period of the wave.

_____ s

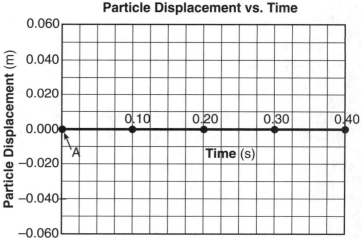

b) On the accompanying grid, starting at point A, sketch a graph of at least one complete cycle of the wave showing its amplitude and period.

c) Calculate the speed of the wave.
[Show all work, including the equation and substitution with units.]

24. The accompanying diagram represents a transverse wave, A, traveling through a uniform medium. On the diagram, draw a wave traveling through the same medium as wave A with twice the amplitude and twice the frequency of wave A.

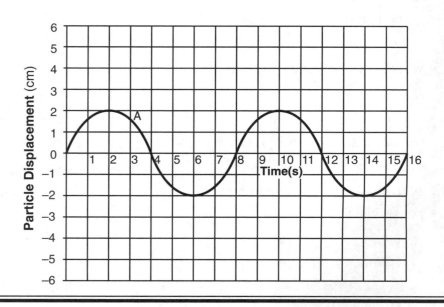

Waves

1. 4   Frequency is defined as the number of waves or pulses occurring per unit time or per second. Since the motor produces 4.0 waves each second, the frequency of the wave is 4.0 Hz.

2. 2   Under Waves, find the equation $v = f\lambda$. Since the question involves light, $v = c$, the speed of light in a vacuum. From the List of Physical Constants on the reference table, $c = 3.00 \times 10^8$ m/s. Substituting into the equation gives 3.00×10^8 m/s $= (5.0 \times 10^{14}$ Hz$)(\lambda)$. $\lambda = 6.0 \times 10^{-7}$ m.

3. 3   Once produced, the frequency of a wave does not change. It depends only upon the source of the wave. Since the period of a wave is the reciprocal of the frequency ($T = 1/f$), it does not change. Under Waves, find the equation $v = f\lambda$. If the speed changes but the frequency remains constant, the wavelength must then change. Phase does not change as a wave enters a new medium.

4. 1   The Electromagnetic Spectrum chart in the reference table shows us that the frequency of green light is greater than that of red light. Under Waves, find the equation $T = 1/f$. From this equation, the larger the frequency, the smaller the period.

5. 1   The period of a wave is defined as the time needed to produce one wave or pulse. The period of this wave is 0.4 s. Under Waves, find the equations $T = 1/f$ and $v = f\lambda$. Use the first equation to find the frequency of the wave: 0.4 s $= 1/f$. Solving for f gives 2.5 Hz. Now use the second equation to solve for the wavelength. Substitution into the equation gives 4.5 m/s $= (2.5$ Hz$)(\lambda)$. Solving for λ gives 1.8 m.

6. 2   The frequency of a wave is the number of waves passing a point in one second. If 20 crests pass a point in one second, twenty waves pass that point in one second. The frequency of the wave is 5 Hz (20 waves/4 seconds). Find the equation $T = 1/f$. The period of this wave is $T = 1/5$ Hz $= 0.2$ s.

7. 1   Radio, television and X-rays are electromagnetic waves (see The Electromagnetic Spectrum). Electromagnetic waves do not need a medium for transmission. Sound waves, being mechanical waves, require a medium for transmission.

8. $v = f\lambda$

$\lambda = \dfrac{v}{f}$

$\lambda = \dfrac{3.00 \times 10^8 \text{ m/s}}{2.2 \times 10^6 \text{ Hz}}$

$\lambda = 1.4 \times 10^2$ m

or 140 m

or 136 m

Explanation: Radio waves are electromagnetic waves similar to light and therefore travel in a vacuum at the speed of light (c) which is found on the List of Physical Constants. Under Waves, find the equation $v = f\lambda$. Use $v = c$ in this equation along with the given frequency to solve for λ.

9. a)

Data Table	
Frequency (Hz)	Wavelength (m)
1.0	6.0
2.0	3.0
3.0	2.0
6.0	1.0

a) Explanation: Under Waves, find the equation $v = f\lambda$. Using $v = 6.0$ m/s and the given frequencies, solve for the wavelengths.

Example: 6.0 m/s = (1.0 Hz)(λ) Solving, $\lambda = 6.0$ m

b and c)

Wavelength vs. Frequency

b) Explanation: Using the wavelength values calculated in question 9a, plot the four data points to within ±0.3 grid space.

c) Explanation: Since the points do not form a straight line, draw the best-fit curve connecting the points.

10. a) $T = \dfrac{1}{f}$

$f = \dfrac{1}{T}$

$f = \dfrac{1}{5.0\,\text{s}}$

$f = 0.20$ Hz

Explanation: The period of a wave is the time needed to generate one complete wave or cycle. The period of the water wave is 5.0 s. Under Waves, find the equation $T = 1/f$. Substitute for T and solve for f.

b) $v = f\lambda$

$v = (0.20\ \text{Hz})(2.0\,\text{m})$ *or*

$v = 0.40$ m/s

$\bar{v} = \dfrac{d}{t}$

$\bar{v} = \dfrac{2.0\ \text{m}}{5.0\ \text{s}}$

$\bar{v} = 0.40$ m/s

Explanation: The wavelength is the distance between two consecutive points on a wave that are phase. For the wave shown, the wavelength is 2.0 m. Under Waves, find the equation $v = f\lambda$. Using the frequency calculated in 10a, calculate v.

or

Under Mechanics, find the equation $\bar{v} = d/t$. The distance a wave travels in a time equal to the period of the wave is the wavelength of the wave. The period of the wave is 5.0 s (see 10a) and the wavelength is 2.0 m. Substitute into the equation and calculate the speed.

Waves – Reflection

Reflection: Reflection is the turning back of a wave when the wave strikes the boundary of a medium in which its speed changes. Part of the wave is turned back into the original medium and part is transmitted into the new medium. The part turned back is the reflected wave and the part transmitted is the refracted wave.

This equation is a mathematical statement of the Law of Reflection, stating that the angle of incidence equals the angle of reflection.

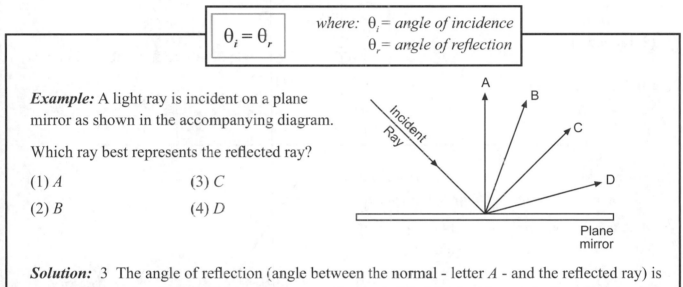

$$\theta_i = \theta_r$$

where: θ_i = angle of incidence
θ_r = angle of reflection

Example: A light ray is incident on a plane mirror as shown in the accompanying diagram.

Which ray best represents the reflected ray?

(1) A (3) C
(2) B (4) D

Solution: 3 The angle of reflection (angle between the normal - letter *A* - and the reflected ray) is equal to the angle of incidence (angle between the normal and the incident ray). Ray *C* shows this relationship.

Reflection – Additional Information:

- The angle of incidence is measured between the incident ray and the normal at the point of incidence. The angle of reflection is measured between the reflected ray and the normal at the point of reflection.

1. The diagram below shows parallel rays of light incident on an irregular surface.

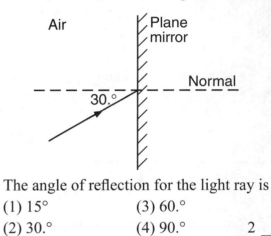

Which phenomenon of light is illustrated by the diagram?

(1) diffraction
(2) refraction
(3) regular reflection
(4) diffuse reflection   1 _____

2. A ray of monochromatic light traveling in air is incident on a plane mirror at an angle of 30.°, as shown in the diagram below.

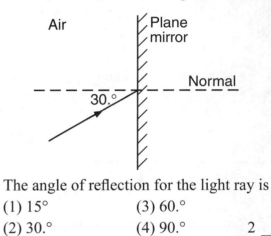

The angle of reflection for the light ray is

(1) 15°   (3) 60.°
(2) 30.°   (4) 90.°   2 _____

Base your answers to questions 3a and b on the information below and accompanying diagram.

A ray of monochromatic light of frequency 5.00×10^{14} hertz is incident on a mirror and reflected, as shown.

3. *a)* Using a protractor and ruler, construct and label the normal to the mirror at the point of incidence on the accompanying diagram.

b) Using a protractor, measure the angle of incidence to the nearest degree.

_____ °

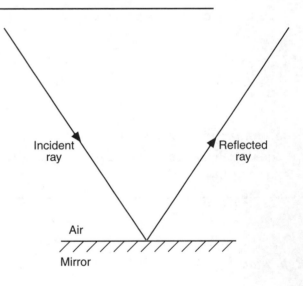

4. The refracted light ray is reflected from the material **X**–air boundary at point P. Using a protractor and straightedge, on the accompanying diagram, draw the reflected ray from point P.

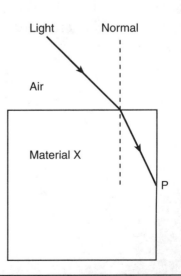

5. Which ray diagram best represents the phenomenon of reflection?

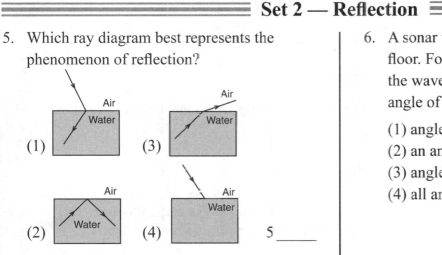

(1)

(3)

(2)

(4)   5 _____

6. A sonar wave is reflected from the ocean floor. For which angles of incidence do the wave's angle of reflection equal its angle of incidence?

(1) angles less than 45°, only
(2) an angle of 45°, only
(3) angles greater than 45°, only
(4) all angles of incidence   6 _____

Base your answers to questions 7a and b on the information and diagram below.

In the diagram, a light ray, R, strikes the boundary of air and water.

R

Air

Water

7. a) Using a protractor, determine the angle of incidence.

   b) Using a protractor and straightedge, draw the reflected ray on the diagram above.

Base your answers to questions 8a and b on the information and diagram below.

Two plane mirrors are positioned perpendicular to each other as shown. A ray of monochromatic red light is incident on mirror 1 at an angle of 55°. This ray is reflected from mirror 1 and then strikes mirror 2.

8. a) Determine the angle at which the ray is incident on mirror 2. _____ °

   b) On the accompanying diagram, use a protractor and a straightedge to draw the ray of light as it is reflected from mirror 2.

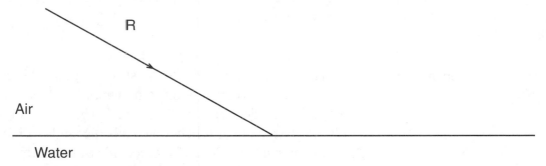

1. **4** Diffuse reflection occurs when light is reflected from a rough or irregular surface. The reflected rays are not parallel to each other.

2. **2** The angle of reflection is equal to the angle of incidence, which is 30°.

3. *a)*

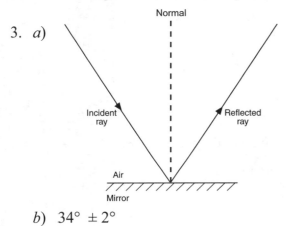

The normal is a line that is perpendicular (90°) to the mirror at the point of incidence

   b) 34° ± 2°

4.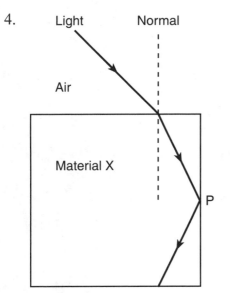

Explanation: In material **X**, the ray striking point *P* is an incident ray. Construct a normal at point *P* in material **X**. Measure the angle between this incident ray and the normal. This is the angle of incidence. The angle of reflection equals the angle of incidence. Construct an equal angle on the other side of the normal and draw the reflected ray at this angle.

Waves – Refraction

Refraction: Refraction is the change in direction of travel or the bending of a wave that occurs when a wave enters obliquely a medium in which its speed changes.

The absolute index of refraction of a medium is the speed of light in a vacuum divided by the speed of light the material medium.

$$n = \frac{c}{v}$$

where: n = *absolute index of refraction*
c = *speed of light in a vacuum*
v = *velocity or speed*

Example: In a certain material, a beam of monochromatic light ($f = 5.09 \times 10^{14}$ hertz) has a speed of 2.25×10^8 meters per second. The material could be

(1) crown glass (2) flint glass (3) glycerol (4) water

Solution: 4 c is the speed of light which is found on the List of Physical Constants. Substitution into the equation and solving for n yields $n = \frac{(3.00) \times 10^8 \text{ m/s}}{(2.25) \times 10^8 \text{ m/s}} = 1.33$. Go to the table of Absolute Indices of Refraction and find which substance has an index of refraction of 1.33. The substance is water.

The equation below is Snell's Law, which relates the absolute indices of refraction and the angles of incidence and refraction as light passes from one medium into another. n_1 is the index of refraction of the medium in which the angle of incidence is measured and n_2 is the index of refraction of the medium in which the angle of refraction is measured.

$$n_1 \sin \theta_1 = n_2 \sin \theta_2$$

where: n = *absolute index of refraction*
θ = *angle*

Example: Base your answer on the accompanying diagram, which represents a light ray traveling from air to Lucite to medium Y and back into air.

The sine of angle x is

(1) 0.333 (3) 0.707
(2) 0.500 (4) 0.886

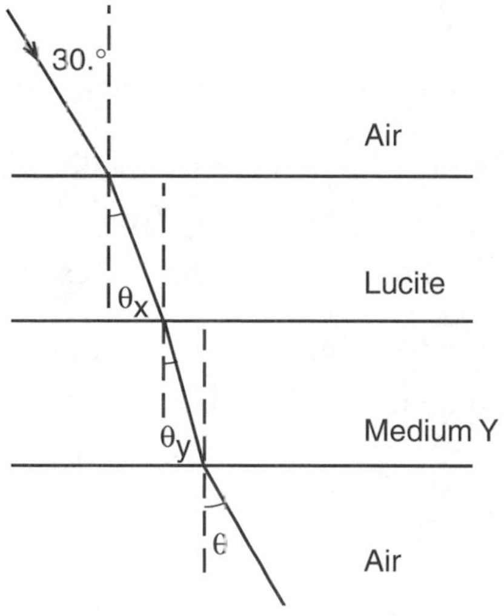

Solution: 1 Use the table of Absolute Indices of Refraction to find the index of refraction of air ($n_1 = 1.00$) and Lucite ($n_2 = 1.50$). Substitution into the equation gives $(1.00)(\sin 30°) = (1.50)(\sin \theta_2)$, where angle θ_2 is angle θ_x. Solving gives $\sin \theta_x = 0.333$.

The ratio $\dfrac{n_2}{n_1}$ is called the relative index of refraction of medium 2 with respect to medium 1. From Snell's Law, it is also equal to $\dfrac{\sin \theta_1}{\sin \theta_2}$.

$$\frac{n_2}{n_1} = \frac{v_1}{v_2} = \frac{\lambda_1}{\lambda_2}$$

where: n = *absolute index of refraction*
v = *velocity or speed*
λ = *wavelength*

Example: A beam of monochromatic light has a wavelength of 5.89×10^{-7} meter in air. Calculate the wavelength of this light in diamond. [Show all work, including the equation and substitution with units.]

Solution:

$$\frac{n_2}{n_1} = \frac{\lambda_1}{\lambda_2}$$

$$\lambda_2 = \frac{n_1 \, \lambda_1}{n_2}$$

$$\lambda_2 = \frac{(1.00)(5.89 \times 10^{-7} \text{ m})}{2.42}$$

$$\lambda_2 = 2.43 \times 10^{-7} \text{ m}$$

Explanation: In this equation, the subscript 1 refers to the original medium, in this case air, and the subscript 2 refers to the new medium, which is diamond. The indices of refraction are found on the table of Absolute Indices of Refraction. Substitution into the first (n_2/n_1) and the third part of the equation (λ_1/λ_2) gives you the correct setup as shown.

Refraction - Additional Information:

- The value of n is found on the Absolute Indices of Refraction table.

- When a wave travels from one medium into another, $\dfrac{n_2}{n_1}$ is the relative index of refraction. The subscripts 1 and 2 refer to the original medium and new medium, respectively. If $\dfrac{n_2}{n_1} > 1$, the speed of the wave decreases as it enters the new medium. If $\dfrac{n_2}{n_1} < 1$, the speed of the wave increases as it enters the new medium.

- The larger the index of refraction of a given medium, the greater the amount of refraction or bending of a ray of light as it enters that medium.

- If a ray of light enters a medium with a smaller index of refraction, its speed increases and it is refracted or bent away from the normal. If a ray of light enters a medium with a larger index of refraction, its speed decreases and it is refracted toward the normal.

- If a ray of light enters a medium with the same index of refraction, there is no change in speed, and therefore there is no refraction or bending.

1. In which way does blue light change as it travels from diamond into crown glass?

    (1) Its frequency decreases.
    (2) Its frequency increases.
    (3) Its speed decreases.
    (4) Its speed increases.   1 _____

2. The diagram below represents straight wave fronts passing from deep water into shallow water, with a change in speed and direction.

    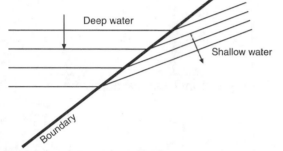

    Which phenomenon is illustrated in the diagram?

    (1) reflection   (3) diffraction
    (2) refraction   (4) interference   2 _____

3. The speed of light ($f = 5.09 \times 10^{14}$ Hz) in a transparent material is 0.75 times its speed in air. The absolute index of refraction of the material is approximately

    (1) 0.75   (3) 2.3
    (2) 1.3   (4) 4.0   3 _____

4. What is the speed of a ray of light ($f = 5.09 \times 10^{14}$ hertz) traveling through a block of sodium chloride?

    (1) 1.54×10^8 m/s
    (2) 1.95×10^8 m/s
    (3) 3.00×10^8 m/s
    (4) 4.62×10^8 m/s   4 _____

5. The diagram below shows a ray of light passing from medium **X** into air.

    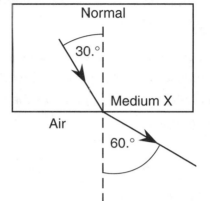

    What is the absolute index of refraction of medium **X**?

    (1) 0.500   (3) 1.73
    (2) 2.00   (4) 0.577   5 _____

6. A ray of light ($f = 5.09 \times 10^{14}$ Hz) traveling in air is incident at an angle of 40.° on an air-crown glass interface as shown below.

    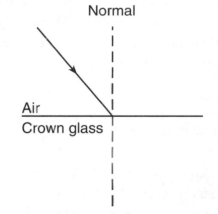

    What is the angle of refraction for this light ray?

    (1) 25°   (3) 40°
    (2) 37°   (4) 78°   6 _____

Base your answers to questions 7a and b on the information and accompanying diagram. A monochromatic beam of yellow light, AB, is incident upon a Lucite block in air at an angle of 33°.

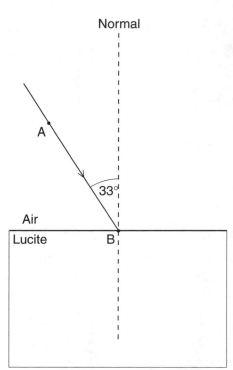

7. a) Calculate the angle of refraction for incident beam AB. [Show all work, including the equation and substitution with units.]

b) Using a straightedge, a protractor, and your answer from question 7a, draw an arrow to represent the path of the refracted beam.

Base your answers to questions 8a and b on the information and accompanying diagram. A ray of light passes from air into a block of transparent material X as shown in the accompanying diagram.

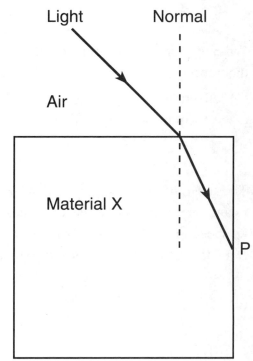

8. a) Measure the angles of incidence and refraction to the nearest degree for this light ray at the air into material X boundary and write your answers in the space below.

angle of incidence _____ °

angle of refraction _____ °

b) Calculate the absolute index of refraction of material X. [Show all work, including the equation and substitution with units.]

Base your answers to questions 9a and b on the accompany diagram, which shows a light ray ($f = 5.09 \times 10^{14}$ Hz) in air, incident on a boundary with fused quartz. At the boundary, part of the light is refracted and part of the light is reflected.

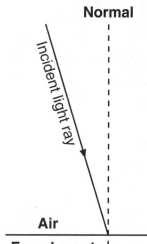

9. a) Using a protractor, measure the angle of incidence of the light ray at the air-fused quartz boundary.

_____ °

b) Calculate the angle of refraction of the incident light ray. [Show all work, including the equation and substitution with units.]

Base your answers to questions 10a and b on the accompanying diagram which shows a ray of monochromatic light ($f = 5.09 \times 10^{14}$ hertz) passing through a flint glass prism.

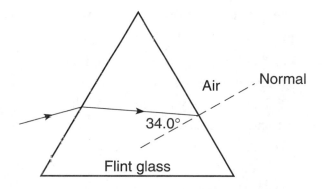

10. a) Calculate the angle of refraction (in degrees) of the light ray as it enters the air from the flint glass prism. [Show all calculations, including the equation and substitution with units.]

b) Using a protractor and a straightedge, construct the refracted light ray in the air on the diagram above.

11. A laser beam is directed at the surface of a smooth, calm pond as represented in the diagram below. Which organisms could be illuminated by the laser light?

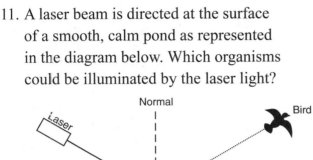

(1) the bird and the fish
(2) the bird and the seaweed
(3) the crab and the seaweed
(4) the crab and the fish   11 _____

12. What happens to the frequency and the speed of an electromagnetic wave as it passes from air into glass?

   (1) The frequency decreases and the speed increases.
   (2) The frequency increases and the speed decreases.
   (3) The frequency remains the same and the speed increases.
   (4) The frequency remains the same and the speed decreases.   12 _____

13. The speed of light in a material is 2.50×10^8 meters per second. What is the absolute index of refraction of the material?

   (1) 1.20   (3) 7.50
   (2) 2.50   (4) 0.833   13 _____

14. Which quantity is equivalent to the product of the absolute index of refraction of water and the speed of light in water?

   (1) wavelength of light in a vacuum
   (2) frequency of light in water
   (3) sine of the angle of incidence
   (4) speed of light in a vacuum   14 _____

15. What is the speed of light ($f = 5.09 \times 10^{14}$ Hz) in flint glass?

   (1) 1.81×10^8 m/s
   (2) 1.97×10^8 m/s
   (3) 3.00×10^8 m/s
   (4) 4.98×10^8 m/s   15 _____

16. The diagram below represents a light ray traveling from air to Lucite to medium Y and back into air.

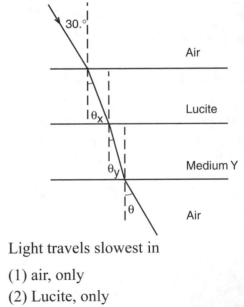

Light travels slowest in

   (1) air, only
   (2) Lucite, only
   (3) medium Y, only
   (4) air, Lucite, and medium Y 16 _____

Base your answers to questions 17a and b on the information and accompanying diagram.

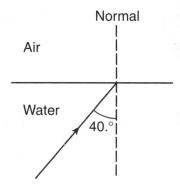

A ray of light of frequency 5.09×10^{14} hertz is incident on a water-air interface as shown in the accompanying diagram.

17. a) Calculate the angle of refraction of the light ray in air. [Show all work, including the equation and substitution with units.]

b) Calculate the speed of the light while in the water. [Show all work, including the equation and substitution with units.]

18. A straight glass rod appears to bend when placed in a beaker of water, as shown in the accompanying diagram. Give an explanation for this phenomenon?

_____

_____

_____

19. A beam of monochromatic light has a wavelength of 5.89×10^{-7} meter in air. Calculate the wavelength of this light in zircon. [Show all work, including the equation and substitution with units.]

Refraction

Answers

Set 1

1. **4** Find the table of Absolute Indices of Refraction in the reference table. The index of diamond is 2.42. That of crown glass is 1.52. Under Waves, find the equation $n = c/v$. Solving for v, $v = c/n$. The value of c is a constant and is found on the List of Physical Constants. Since the light is entering a medium with a lower index of refraction, as shown by the equation for v, the speed of the light will increase.

2. **2** Refraction is defined as the change in direction of travel of a wave as it enters obliquely (at an angle of incidence other than $0°$) a medium in which its speed changes.

3. **2** Under Waves, find the equation $n = c/v$. Since the absolute index of refraction of air is 1.00, the speed of light in air is equal to the speed of light in a vacuum (c). In the question, the speed of light in the transparent material (v) is $0.75\ c$. Substitution into the equation gives
$$n = \frac{c}{(0.75\ c)} = \frac{1}{0.75} = 1.3.$$

4. **2** Under Waves, find the equation $n = c/v$. The value of the speed of light in a vacuum is given on the List of Physical Constants and the index of refraction of sodium chloride is found on the table of Absolute Indices of Refraction in the reference table. Substituting these values into the equation gives $1.54 = (3.00 \times 10^8\ \text{m/s})/v$. Solving, $v = 1.95 \times 10^8\ \text{m/s}$.

5. **3** Find the equation $n_1 \sin \theta_1 = n_2 \sin \theta_2$ (Snell's Law). The absolute index of refraction of air is found on the table of Absolute Indices of Refraction. Substitution gives $n_1(\sin 30°) = (1.00)(\sin 60°)$, where n_1 is the index of refraction of medium **X**. Solving, $n_1 = 1.73$.

6. **1** Find the equation $n_1 \sin \theta_1 = n_2 \sin \theta_2$. The absolute index of refraction of air and crown glass are found on the table of Absolute Indices of Refraction. Substitution gives $(1.00)(\sin 40°) = (1.52)(\sin \theta_2)$. Solving, $\sin \theta_2 = 0.422$ and $\theta_2 = 25°$.

7. *a)* $n_1 \sin \theta_1 = n_2 \sin \theta_2 \Rightarrow (1.00)(\sin 33°) = (1.50) \sin \theta_2 \quad \sin \theta_2 = 0.363 \quad \theta_2 = 21°$

Explanation: Under Waves, find the equation $n_1 \sin \theta_1 = n_2 \sin \theta_2$ (Snell's Law). Using the table of Absolute Indices of Refraction in the reference table, find the index of refraction of air (1.00) and Lucite (1.50). Using $n_1 = 1.00$, $\theta_1 = 33°$ and $n_2 = 1.50$, substitute into the equation and solve for θ_2, the angle of refraction in Lucite.

b)

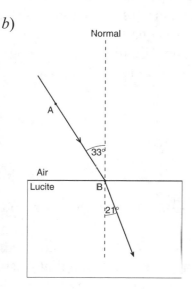

Explanation: From question 7*a*, the angle of refraction in Lucite is 21°. This is the angle between the refracted ray in Lucite and the normal.

8. *a)* The angle of incidence is 45° (±2°).
The angle of refraction is 26° (±2°).

Explanation: The angle of incidence is the angle between the incident ray and the normal. The angle of refraction is the angle between the refracted ray and the normal. They are not measured with respect to the boundary.

b)
$$n_1 \sin \theta_1 = n_2 \sin \theta_2$$
$$n_2 = \frac{(1.00)(\sin 45°)}{\sin 26°}$$
$$n_2 = 1.61$$

Explanation: Under Waves, find the equation $n_1 \sin \theta_1 = n_2 \sin \theta_2$. Find the index of refraction of air on the Absolute Indices. Using $n_1 = 1.00$, $\theta_1 =$ your measured angle of incidence in question 8*a* and $\theta_2 =$ your measured angle of refraction in question 8*a*. Solve for n_2.

9. *a)* Answer: 17°

Explanation: The angle of incidence is the angle between the incident ray and the normal to the surface at the point of incidence. It must be within ±2°.

b) $n_1 \sin \theta_1 = n_2 \sin \theta_2$

$\sin \theta_2 = \dfrac{n_1 \sin \theta_1}{n_2}$

$\sin \theta_2 = \dfrac{1.00 \sin 17°}{1.46}$

$\theta_2 = 12°\ or\ 11.6°$

Explanation: Under Waves, find the equation $n_1 \sin \theta_1 = n_2 \sin \theta_2$ (Snell's Law). Values of n_1 (absolute index of refraction of air) and n_2 (absolute index of refraction of fused quartz) are found on the table of Absolute Indices of Refraction. Use these values along with the angle of incidence (θ_1) measured in question 9a to calculate the angle refraction (θ_2).

10. *a)* $n_1 \sin \theta_1 = n_2 \sin \theta_2$ $\sin \theta_2 = \dfrac{n_1 \sin \theta_1}{n_2}$ $\sin \theta_2 = \dfrac{(1.66)(\sin 34.0°)}{(1.00)}$ $\theta_2 = 68.2°\ \textbf{\textit{or}}\ 68°$

Explanation: Under Waves, find the equation $n_1 \sin \theta_1 = n_2 \sin \theta_2$. Use the table of Absolute Indices of Refraction in the reference table to find the indices of refraction of air (1.00) and flint glass (1.66), n_1 and n_2, respectively. Angle θ_1 is 34.0°. Substitute into the equation and calculate θ_2.

b)

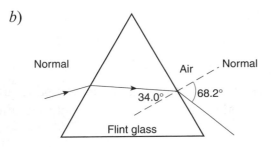

Explanation: The angle of refraction is defined as the angle between the normal and the refracted ray. Measure the angle calculated in question 10a with respect to the normal in air. Draw in the refracted ray at this angle.

Modern Physics deals with the structure and properties of the atom, in particular, the quantum model of the atom.

Photon energy: A photon is a bundle of electromagnetic radiation.

The energy of a photon or "particle" of light is equal to the product of Planck's constant and the frequency of the light. From the wave speed equation $f = c/\lambda$ (see Waves), where c, the speed of light in a vacuum, has been substituted for v. Therefore $E_{photon} = hc/\lambda$.

$$E_{photon} = hf = \frac{hc}{\lambda}$$

where: $E = energy$
$h = Planck's\ constant$
$f = frequency$
$c = speed\ of\ light\ in\ a\ vacuum$
$\lambda = wavelength$

Example: Light of wavelength 5.0×10^{-7} meter consists of photons having an energy of

(1) 1.1×10^{-48} J   (2) 1.3×10^{-27} J   (3) 4.0×10^{-19} J   (4) 1.7×10^{-5} J

Solution: 3   For this problem one must use the equation $E_{photon} = hc/\lambda$. The values of h and c are found on the List of Physical Constants. Substitution into the equation

gives: $E_{photon} = \dfrac{(6.63 \times 10^{-34}\ J \bullet s)(3.00 \times 10^{8}\ m/s)}{(5.0 \times 10^{-7}\ m)} = 4.0 \times 10^{-19}\ J.$

Photon Energy – Additional Information:

- As the frequency of the radiation increases, or the wavelength decreases, the particle nature of the radiation increases.

- Energy is absorbed or emitted by atoms in the form of photons.

- The value of c and h is found on the List of Physical Constants.

1. Wave-particle duality is most apparent in analyzing the motion of

   (1) a baseball   (3) a galaxy
   (2) a space shuttle   (4) an electron 1 _____

2. Which graph best represents the relationship between photon energy and photon frequency?

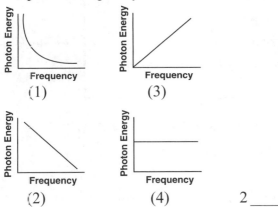

        (1)   (3)

        (2)   (4)   2 _____

3. The energy of a photon is inversely proportional to its

   (1) wavelength   (3) frequency
   (2) speed   (4) phase   3 _____

4. Compared to a photon of red light, a photon of blue light has a

   (1) greater energy
   (2) longer wavelength
   (3) smaller momentum
   (4) lower frequency   4 _____

5. What is the energy of a photon with a frequency of 5.00×10^{14} hertz?

   (1) 3.32 eV
   (2) 3.20×10^{-6} eV
   (3) 3.00×10^{48} J
   (4) 3.32×10^{-19} J   5 _____

6. A photon of light traveling through space with a wavelength of 6.0×10^{-7} meter has an energy of

   (1) 4.0×10^{-40} J
   (2) 3.3×10^{-19} J
   (3) 5.4×10^{10} J
   (4) 5.0×10^{14} J   6 _____

7. The graph below represents the relationship between the energy and the frequency of photons.

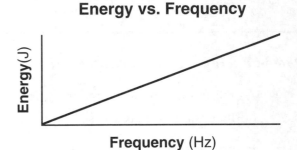

Energy vs. Frequency

The slope of the graph would be
   (1) 6.63×10^{-34} J•s
   (2) 6.67×10^{-11} N•m²/kg²
   (3) 1.60×10^{-19} J
   (4) 1.60×10^{-19} C   7 _____

Base your answers to questions 8a and b on the information and diagram below. The diagram shows the collision of an incident photon having a frequency of 2.00×10^{19} hertz with an electron initially at rest.

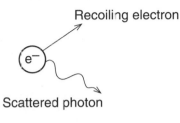

Before collision   **After collision**

Incident photon   Electron   Scattered photon
                   at rest

8. *a*) Calculate the initial energy of the photon. [Show all calculations, including the equation and substitution with units.]

    b) What is the total energy of the two-particle system after the collision?

9. Determine the color of a photon with an energy of 3.20×10^{-19} J.

Base your answers to questions 10a and b on the information below.

A photon with a frequency of 5.02×10^{14} hertz is absorbed by an excited hydrogen atom. This causes the electron to be ejected from the atom, forming an ion.

10. *a*) Calculate the energy of this photon in joules. [Show all work, including the equation and substitution with units.]

    b) Determine the energy of this photon in electronvolts. _____ eV

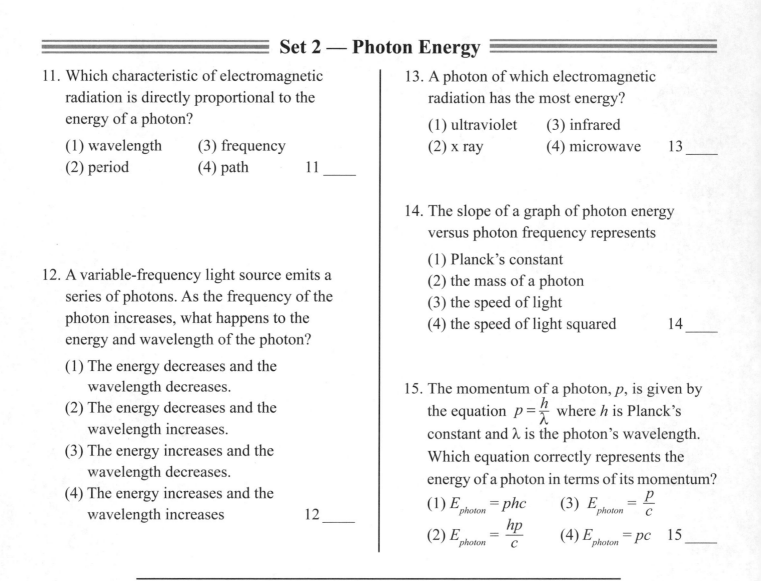

11. Which characteristic of electromagnetic radiation is directly proportional to the energy of a photon?

    (1) wavelength   (3) frequency
    (2) period   (4) path   11 _____

12. A variable-frequency light source emits a series of photons. As the frequency of the photon increases, what happens to the energy and wavelength of the photon?

    (1) The energy decreases and the wavelength decreases.
    (2) The energy decreases and the wavelength increases.
    (3) The energy increases and the wavelength decreases.
    (4) The energy increases and the wavelength increases   12 _____

13. A photon of which electromagnetic radiation has the most energy?

    (1) ultraviolet   (3) infrared
    (2) x ray   (4) microwave   13 _____

14. The slope of a graph of photon energy versus photon frequency represents

    (1) Planck's constant
    (2) the mass of a photon
    (3) the speed of light
    (4) the speed of light squared   14 _____

15. The momentum of a photon, p, is given by the equation $p = \dfrac{h}{\lambda}$ where h is Planck's constant and λ is the photon's wavelength. Which equation correctly represents the energy of a photon in terms of its momentum?

    (1) $E_{photon} = phc$ (3) $E_{photon} = \dfrac{p}{c}$
    (2) $E_{photon} = \dfrac{hp}{c}$ (4) $E_{photon} = pc$ 15 _____

16. Calculate the wavelength of a photon having 3.26×10^{-19} joule of energy. [Show all work, including the equation and substitution with units.]

17. A photon has a wavelength of 9.00×10^{-10} meter. Calculate the energy of this photon in joules. [Show all work, including the equation and substitution with units.]

The light of the "alpha line" in the Balmer series of the hydrogen spectrum has a wavelength of 6.58×10^{-7} meter.

18. *a*) Calculate the energy of an "alpha line" photon in joules. [Show all work, including the equation and substitution with units.]

   b) What is the energy of an "alpha line" photon in electronvolts? _____ eV

19. Exposure to ultraviolet radiation can damage skin. Exposure to visible light does not damage skin. State one possible reason for this difference.

   _____

   _____

   _____

   Base your answers to questions 20*a*, *b*, and *c* on the information below.

   A photon with a frequency of 5.48×10^{14} hertz is emitted when an electron in a mercury atom falls to a lower energy level.

20. *a*) Identify the color of light associated with this photon. _____

   b) Calculate the energy of this photon in joules. [Show all work, including the equation and substitution with units.]

   c) Determine the energy of this photon in electronvolts. _____ eV

1. 4   The wave properties of matter particles become important and noticeable only for particles of extremely small mass. Since the electron has the smallest mass of the choices given, it will most likely show wave-particle duality.

2. 3   Under Modern Physics, find the equation $E_{photon} = hf$. This shows that the energy of the photon varies directly with the frequency.

3. 1   Find the equation $E_{photon} = hf = hc/\lambda$. This equation shows that the energy of a photon varies directly with frequency and inversely with wavelength.

4. 1   Locate the equation $E_{photon} = hf$. This indicates that the energy of a photon varies directly with the frequency of the light. From The Electromagnetic Spectrum chart, note that blue light has a higher frequency than red light. Therefore, a photon of blue light has greater energy than a photon of red light.

5. 4   Under Modern Physics, find the equation $E_{photon} = hf$. The value of h (Planck's constant) is found on the List of Physical Constants. Substitution gives $E_{photon} = (6.63 \times 10^{-34}$ J•s$)(5.00 \times 10^{14}$ Hz$)$. Solving, $E_{photon} = 3.32 \times 10^{-19}$ J.

6. 2   Find the equation $E_{photon} = hf = hc/\lambda$. The values of h and c are found on the List of Physical Constants. Substitution gives $E_{photon} = (6.63 \times 10^{-34}$ J•s$)(3.00 \times 10^8$ m/s$)/(6.0 \times 10^{-7}$ m$)$. Solving, $E_{photon} = 3.3 \times 10^{-19}$ J.

7. 1   The slope of a graph is defined as the change in y (energy) divided by the change in x (frequency). For this graph, slope = $\Delta E/\Delta f$. Under Modern Physics, find the equation $E_{photon} = hf$. Solving for the ratio of energy to frequency gives $E_{photon}/f = h$, which is Planck's constant, found on the List of Physical Constants.

8. *a)* $E_{photon} = hf \Rightarrow E_{photon} = (6.63 \times 10^{-34}$ J•s$)(2.00 \times 10^{19}$ Hz$) = 1.33 \times 10^{-14}$ J.

   Explanation: Under Modern Physics, find the equation $E_{photon} = hf$. On the List of Physical Constants, find the value of h (Planck's constant). Substitute into the equation for h and f and calculate the photon energy.

   b)   The energy of the system after the collision is 1.3×10^{-14} J

   Explanation: During the collision between the photon and the electron, energy must be conserved. Since the electron was at rest before the collision, the total energy before the collision was that of the photon before the collision.

9.   Orange

Explanation: Under Modern Physics, find the equation $E_{photon} = hf$. The value of h is found on the List of Physical Constants. Substitution gives 3.20×10^{-19} J $= (6.63 \times 10^{-34}$ J•s$)(f)$. Solving, $f = 4.83 \times 10^{14}$ Hz. Referring to The Electromagnetic Spectrum chart , this frequency is in the orange part of the visible spectrum.

10. a) $E_{photon} = hf$

$E_{photon} = (6.63 \times 10^{-34}$ J•s$)(5.02 \times 10^{14}$ Hz$)$

$E_{photon} = 3.33 \times 10^{-19}$ J

Explanation: Under Modern Physics, find the equation $E_{photon} = hf$. The value of h (Planck's constant) is found on the List of Physical Constants. Substitute the value of h and the given frequency into the equation and solve for the energy of the photon.

b)   Answer: 2.08 eV

Explanation: On the List of Physical Constants, you will find that 1 electronvolt (eV) $= 1.60 \times 10^{-19}$ J. Use this as a conversion factor to convert the answer in question 10a from J to eV.

3.33×10^{-19} J $\times \dfrac{1\ eV}{1.60 \times 10^{-19}\ J} = 2.08$ eV

Modern Physics – Energy Level Transitions

Energy Level Transitions: Electrons in atoms may move from one energy level to another by absorbing or radiating energy in the form of photons.

The energy absorbed or radiated by an atom as an electron moves from one energy level to another in the atom is exactly equal to the difference between the energy of the initial level and the energy of the final level.

$$E_{photon} = E_i - E_f \qquad where: \ E = energy$$

Example: A hydrogen atom with an electron initially in the $n = 2$ level is excited further until the electron is in the $n = 4$ level. This energy level change occurs because the atom has

(1) absorbed a 0.85-eV photon

(3) absorbed a 2.55-eV photon

(2) emitted a 0.85-eV photon

(4) emitted a 2.55-eV photon

Solution: 3 Using the Energy Level Diagram for Hydrogen and substituting into the equation gives $E_{photon} = (-3.40 \ eV) - (-0.85 \ eV) = -2.55 \ eV$. Since the electron is moving to a higher energy level, this photon is absorbed by the atom, as indicated by the negative sign.

Energy Level Transitions – Additional Information:

- This equation is used in conjunction with the Energy Level Diagrams of the Hydrogen atom and the Mercury atom to determine the energy of a photon as the atom undergoes an energy level transition.

- E_i and E_f are the initial energy level and final energy level of the atom, respectively, taken from the Energy Level Diagrams.

- The relationship shown between the electronvolt (eV) and the joule (J) on the List of Physical Constants may be used as a conversion factor between these energy units.

1. What is the minimum energy needed to ionize a hydrogen atom in the $n = 2$ energy state?

   (1) 13.6 eV   (3) 3.40 eV

   (2) 10.2 eV   (4) 1.89 eV   1 _____

2. An electron in a mercury atom drops from energy level f to energy level c by emitting a photon having an energy of

   (1) 8.20 eV   (3) 2.84 eV

   (2) 5.52 eV   (4) 2.68 eV   2 _____

3. A mercury atom in the ground state absorbs 20.00 electronvolts of energy and is ionized by losing an electron. How much kinetic energy does this electron have after the ionization?

   (1) 6.40 eV   (3) 10.38 eV

   (2) 9.62 eV   (4) 13.60 eV   3 _____

4. Which type of photon is emitted when an electron in a hydrogen atom drops from the $n = 2$ to the $n = 1$ energy level?

   (1) ultraviolet   (3) infrared

   (2) visible light   (4) radio wave   4 _____

Base your answers to questions 5a, b, c and d on the Energy Level Diagram for Hydrogen in the Reference Tables for Physical Settings/Physics.

5. a) Determine the energy, in electronvolts, of a photon emitted by an electron as it moves from the $n = 6$ to the $n = 2$ energy level in a hydrogen atom.

   b) Convert the energy of the photon to joules.

   c) Calculate the frequency of the emitted photon. [Show all work, including the equation and substitution with units.]

   d) Is this the only energy and/or frequency that an electron in the n = 6 energy level of a hydrogen atom could emit? Explain your answer.

_____

_____

6. Excited hydrogen atoms are all in the $n = 3$ state. How many different photon energies could possibly be emitted as these atoms return to the ground state?

   (1) 1   (2) 2   (3) 3   (4) 4   6 _____

Base your answers to question 7 a and b on the accompanying diagram, which shows some energy levels for an atom of an unknown substance.

7. a) Determine the minimum energy necessary for an electron to change from the B energy level to the F energy level.

   b) An electron in this atom in the C energy level absorbs a 15.00 eV photon and is ionized by losing this electron. Calculate the kinetic energy of this electron after the ionization.

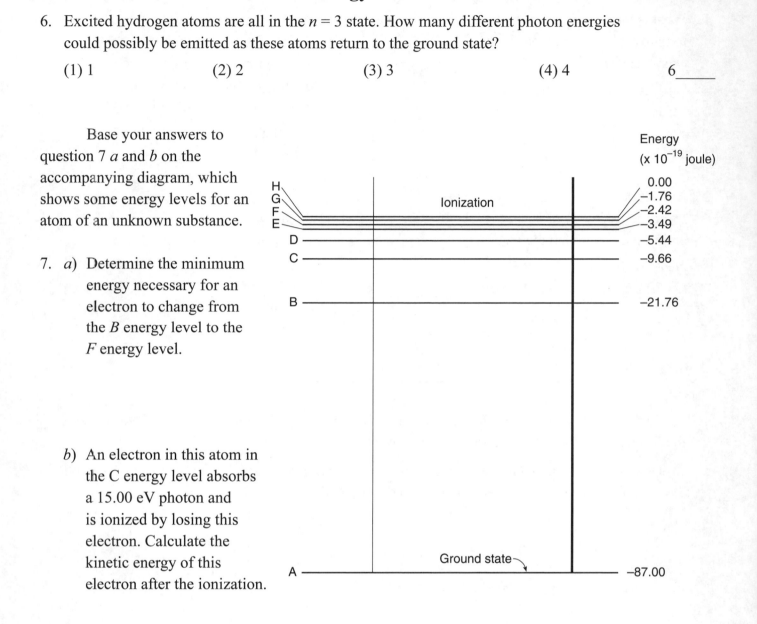

8. Explain why a hydrogen atom in the ground state can absorb a 10.2-electronvolt photon, but can not absorb an 11.0-electronvolt photon.

   _____

   _____

Modern Physics

Energy Level Transitions
Answers – Set 1

1. **3** Referring to the Energy Level Diagram for the hydrogen atom, to ionize the hydrogen atom, it must be excited to the $n = \infty$ energy level. Under Modern Physics in the reference table, find the equation $E_{photon} = E_i - E_f$. $E_{photon} = -3.40$ eV $- 0.00$ eV $= -3.40$ eV. The negative sign indicates that energy is being added to the atom.

2. **3** Find the equation $E_{photon} = E_i - E_f$. Using the Energy Level Diagram for Mercury, $E_{photon} = (-2.68$ eV$) - (-5.52$ eV$) = 2.84$ eV. The positive energy value indicates that energy is released during this transition.

3. **2** Find the Energy Level Diagrams. Referring to that of Mercury, the ionization energy is 10.38 eV, the difference between the ground state (Level a) and the ionization state (Level j). The excess energy carried by the photon will be carried away by the electron as kinetic energy (20.00 eV − 10.38 eV = 9.62 eV).

4. **1** Under Modern Physics, find the equation $E_{photon} = E_i - E_f$. Using the Energy Level Diagram for Hydrogen, $E_{photon} = (-3.40$ eV$) - (-13.60$ eV$) = -10.20$ eV. The negative sign indicates that the photon is emitted. Now find the equation $E_{photon} = hf$ under Modern Physics. The value of h is given on the List of Physical Constants. In this equation, energy must be in joules (J). On the List of Physical Constants, 1 eV $= 1.60 \times 10^{-19}$ J.

   Use this relationship to convert eV to J: 10.20 eV $\times \dfrac{(1.60 \times 10^{-19} \text{ J})}{(1 \text{ eV})} = 1.63 \times 10^{-18}$ J. Substitution

   into the equation gives 1.63×10^{-18} J $= (6.63 \times 10^{-34}$ J•s$)(f)$. Solving, $f = 2.46 \times 10^{15}$ Hz, which according to the Electromagnetic Spectrum chart would be ultraviolet radiation.

5. *a)* Answer: 3.02 eV

   Explanation: Under Modern Physics, find the equation $E_{photon} = E_i - E_f$. Using the Energy Level Diagram for Hydrogen, $E_{photon} = (-0.38$ eV$) - (-3.40$ eV$) = 3.02$ eV. The positive energy value indicates that energy is released during this transaction.

   b) Answer: 4.83×10^{-19} J

   Explanation: On the List of Physical Constants, 1 eV $= 1.60 \times 10^{-19}$ J. Using this relationship to convert eV to J: 3.02 eV $\times \dfrac{(1.60 \times 10^{-19} \text{ J})}{(1 \text{ eV})} = 4.83 \times 10^{-19}$ J

   c) $E = hf$

   $f = \dfrac{E}{h}$

   Explanation: Under Modern Physics find the equation $E_{photon} = hf$. The value of h is given in the List of Physical Constants. Substitution into the equation gives the setup and answer shown.

   $f = \dfrac{4.83 \times 10^{-19} \text{ J}}{6.63 \times 10^{-34} \text{ J•s}}$

   $f = 7.29 \times 10^{14}$ Hz

   d) Acceptable responses include, but are not limited to:
   — No, the $n = 6$ level can return to any of the 5 lower energy levels.
   — No, the electron can drop to many different energy levels.
   — The electron can fall from $n = 6$ to any other level between $n = 5$ and $n = 1$.
   — $6 \rightarrow 5$ $6 \rightarrow 4$ $6 \rightarrow 3$ $6 \rightarrow 1$

Modern Physics – Mass-Energy Equivalence

Mass-Energy Equivalence: This is Einstein's famous mass-energy equation, which indicates that mass may be converted to energy and vice versa.

> The energy equivalent of a particle of mass is the product of the mass of the particle and the square of the speed of light in a vacuum.

$$E = mc^2$$

where: E = energy

m = mass

c = speed of light in a vacuum

Example: The energy produced by the complete conversion of 2.0×10^{-5} kilograms of mass into energy is

(1) 1.8 TJ   (2) 6.0 GJ   (3) 1.8 MJ   (4) 6.0 kJ

Solution: 1 $E = mc^2$ is Einstein's equation, which relates mass and energy. The value of c is found on the List of Physical Constants. Substitution gives $E = (2.0 \times 10^{-5}\text{ kg})(3.00 \times 10^8\text{ m/s})^2$. Solving, $E = 1.8 \times 10^{12}$ J. Using the table of Prefixes for Powers of 10, the prefix for 10^{12} is tera (T). The answer may then be expressed as 1.8 TJ.

Mass-Energy Equivalence – Additional Information:

- The value of c is found on the List of Physical Constants.

- The value of c is very large. When squared in this equation, it indicates that a small amount of matter, when changed to energy, produces a large amount of energy.

- The relationship shown between the electronvolt (eV) and the joule (J) on the List of Physical Constants may be used as a conversion factor between these energy units.

- The relationship shown between the universal mass unit (u) and millions of electronvolts (MeV) on the List of Physical Constants may be used as a conversion factor between mass and energy on the atomic scale.

Base your answer to question 1 on the cartoon below and your knowledge of physics.

1. In the cartoon, Einstein is contemplating the equation for the principle that

   (1) the fundamental source of all energy is the conversion of mass into energy
   (2) energy is emitted or absorbed in discrete packets called photons
   (3) mass always travels at the speed of light in a vacuum
   (4) the energy of a photon is proportional to its frequency   1 _____

2. How much energy would be generated if a 1.0×10^{-3}-kilogram mass were completely converted to energy?

   (1) 9.3×10^{-1} MeV
   (2) 9.3×10^{2} MeV
   (3) 9.0×10^{13} J
   (4) 9.0×10^{16} J   2 _____

3. The energy equivalent of the rest mass of an electron is approximately

   (1) 5.1×10^{5} J
   (2) 8.2×10^{-14} J
   (3) 2.7×10^{-22} J
   (4) 8.5×10^{-28} J   3 _____

4. After a uranium nucleus emits an alpha particle, the total mass of the new nucleus and the alpha particle is less than the mass of the original uranium nucleus. Explain what happens to the missing mass.

   _____

   _____

   _____

5. Which graph best represents the relationship between energy and mass when matter is converted into energy?

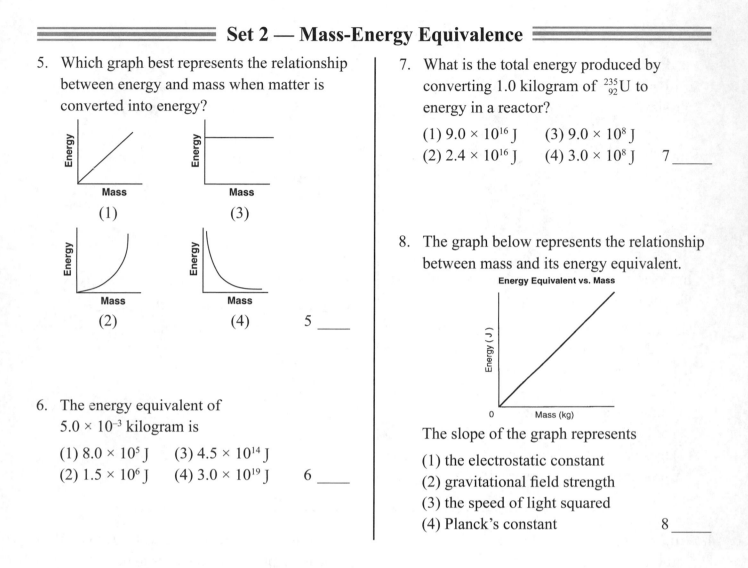

(1)

(3)

(2)

(4)   5 ____

6. The energy equivalent of 5.0×10^{-3} kilogram is

    (1) 8.0×10^5 J   (3) 4.5×10^{14} J
    (2) 1.5×10^6 J   (4) 3.0×10^{19} J   6 ____

7. What is the total energy produced by converting 1.0 kilogram of $^{235}_{92}$U to energy in a reactor?

    (1) 9.0×10^{16} J   (3) 9.0×10^8 J
    (2) 2.4×10^{16} J   (4) 3.0×10^8 J   7 ____

8. The graph below represents the relationship between mass and its energy equivalent.

    The slope of the graph represents

    (1) the electrostatic constant
    (2) gravitational field strength
    (3) the speed of light squared
    (4) Planck's constant   8 ____

9. How much energy, in megaelectronvolts, is produced when 0.250 universal mass units of matter is completely converted into energy? [Show all work, including the equation and substitution with units.]

    _____ MeV

10. Explain how it is possible for a colliding proton and antiproton to produce a particle with six times the mass of either.

    _____

    _____

    _____

1. 1   The equation on the chalk board that Einstein is contemplating resembles the equation $E = mc^2$, which is the equation that says mass may be converted into energy and vice versa with exact equivalence.

2. 3   Find the equation $E = mc^2$. The value of c is found on the List of Physical Constants. Substitution gives $E = (1.0 \times 10^{-3}\ kg)(3.00 \times 10^8\ m/s)^2$. Solving, $E = 9.0 \times 10^{13}\ J$.

3. 2   Under Modern Physics, find the equation $E = mc^2$. In the List of Physical Constants, find the rest mass of the electron (m) and the speed of light (c). Substitution of these values into the equation gives $E = (9.11 \times 10^{-31}\ kg)(3.00 \times 10^8\ m/s)^2$.
Solving gives an energy of $8.2 \times 10^{-14}\ J$.

4.   The missing mass has been converted to energy according to the equation $E = mc^2$.

Geometry and Trigonometry: Geometry is the branch of mathematics concerned with the size, shape, properties and relative position of mathematical figures. Trigonometry is the branch of mathematics that deals with the relationships between angles and the lengths of the sides of a triangle. For a right triangle, these relationships are the Pythagorean Theorem, sin, cos, and tan functions.

Area is the measure of the bounded region on a plane or surface of a solid.

Rectangle $A = bh$	where: A = area b = base h = height

Example: A rectangular nichrome wire heating element measures 0.50 mm by 4.0 mm. Calculate the cross-sectional area of the wire in m².

Solution: 2.0×10^{-6} m²

Explanation: To determine the cross-sectional area in m², the dimensions of the wire must be converted to m. From the Prefixes for Powers of 10 table, milli- (m) = 10^{-3}. Using this as a conversion factor, 0.50 mm × (10^{-3} m)/(1 mm) = 0.50×10^{-3} m = 5.0×10^{-4} m. In a similar fashion, 4.0 mm = 4.0×10^{-3} m.

Substitution and solving gives $A = (5.0 \times 10^{-4}$ m$)(4.0 \times 10^{-3}$ m$) = 2.0 \times 10^{-6}$ m².

Additional Information:

- On a graph of speed *vs.* time where the speed is constant, the area of a rectangle equals the distance traveled in a given interval of time.

- This equation can be used to calculate the cross-sectional area of a flat or rectangular conductor such as a heating element in a toaster.

Triangle $A = \frac{1}{2}bh$	where: A = area b = base h = height

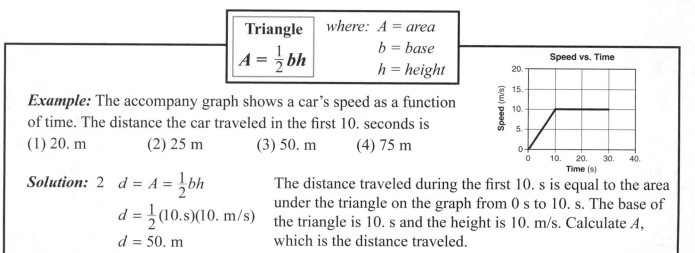

Example: The accompany graph shows a car's speed as a function of time. The distance the car traveled in the first 10. seconds is

(1) 20. m   (2) 25 m   (3) 50. m   (4) 75 m

Solution: 2 $d = A = \frac{1}{2}bh$

$d = \frac{1}{2}(10.\text{s})(10.\text{ m/s})$

$d = 50.\text{ m}$

The distance traveled during the first 10. s is equal to the area under the triangle on the graph from 0 s to 10. s. The base of the triangle is 10. s and the height is 10. m/s. Calculate A, which is the distance traveled.

Additional Information:

- On a graph of speed *vs.* time where the speed is changing, the area of a triangle equals the distance traveled in a given interval of time.

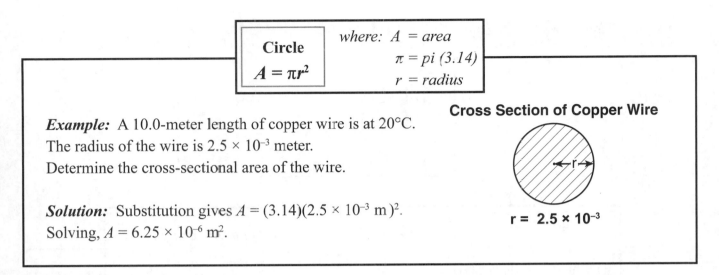

Circle $A = \pi r^2$	where: A = area π = pi (3.14) r = radius

Cross Section of Copper Wire

Example: A 10.0-meter length of copper wire is at 20°C. The radius of the wire is 2.5×10^{-3} meter. Determine the cross-sectional area of the wire.

$r = 2.5 \times 10^{-3}$

Solution: Substitution gives $A = (3.14)(2.5 \times 10^{-3} \text{ m})^2$. Solving, $A = 6.25 \times 10^{-6} \text{ m}^2$.

Additional Information:

- This equation is used to calculate the cross-sectional area of a round conductor.

Circumference is the distance around a closed curve.

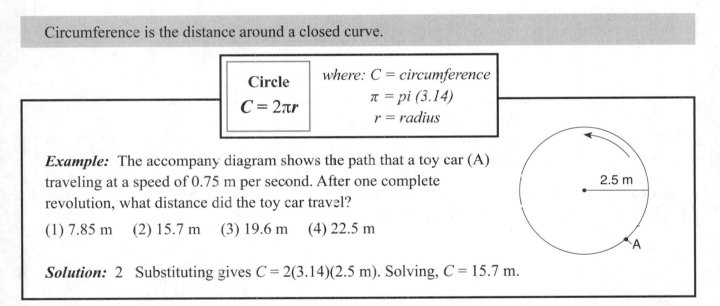

Circle $C = 2\pi r$	where: C = circumference π = pi (3.14) r = radius

Example: The accompany diagram shows the path that a toy car (A) traveling at a speed of 0.75 m per second. After one complete revolution, what distance did the toy car travel?

(1) 7.85 m   (2) 15.7 m   (3) 19.6 m   (4) 22.5 m

2.5 m

Solution: 2   Substituting gives $C = 2(3.14)(2.5 \text{ m})$. Solving, $C = 15.7$ m.

Additional Information:

- This equation is used to calculate the distance traveled in one lap around a circular path.

A right triangle is one in which one angle is equal to 90°.

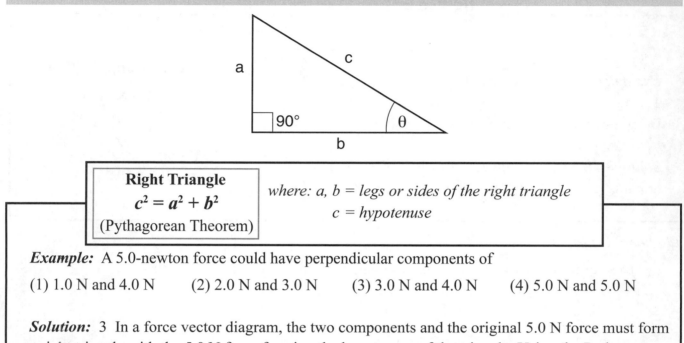

$$\begin{array}{c} \textbf{Right Triangle} \\ \boldsymbol{c^2 = a^2 + b^2} \\ \text{(Pythagorean Theorem)} \end{array}$$

where: a, b = legs or sides of the right triangle
c = hypotenuse

Example: A 5.0-newton force could have perpendicular components of

(1) 1.0 N and 4.0 N   (2) 2.0 N and 3.0 N   (3) 3.0 N and 4.0 N   (4) 5.0 N and 5.0 N

Solution: 3 In a force vector diagram, the two components and the original 5.0 N force must form a right triangle with the 5.0 N force forming the hypotenuse of the triangle. Using the Pythagorean Theorem, the 3.0 N and 4.0 N components are the only components that fit this condition: $c^2 = (3.0 \text{ N})^2 + (4.0 \text{ N})^2$ and $c = 5.0$ N.

$$\begin{array}{c} \textbf{Right Triangle} \\ \boldsymbol{\sin \theta = a/c} \end{array}$$

where: a = side opposite θ
c = hypotenuse

Example: Referring to the diagram, using trigonometry, calculate the vertical component of the 24 N force.

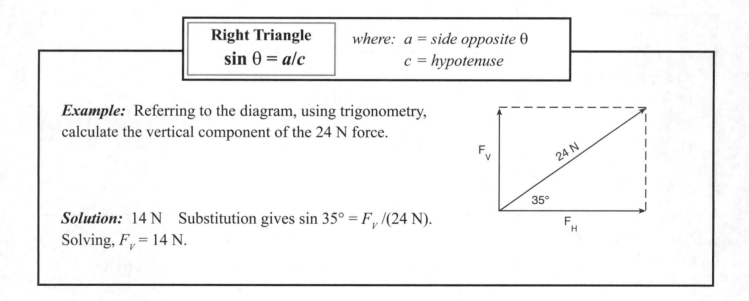

Solution: 14 N   Substitution gives $\sin 35° = F_V /(24 \text{ N})$. Solving, $F_V = 14$ N.

Right Triangles continued on next page

Right Triangle	*where:* b = *side adjacent to* θ
$\cos \theta = b/c$	c = *hypotenuse*

Example: Referring to the diagram, using trigonometry, calculate the horizontal component of the 24 N force.

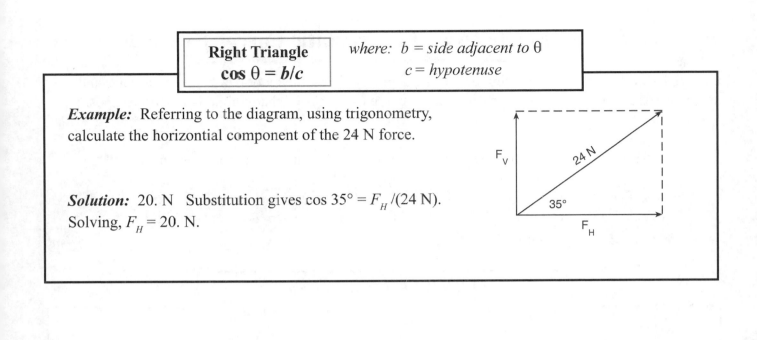

Solution: 20. N   Substitution gives $\cos 35° = F_H /(24\ \text{N})$. Solving, $F_H = 20.$ N.

Right Triangle	*where:* a = *side opposite* θ
$\tan \theta = a/b$	b = *side adjacent to* θ

Example: Referring to the diagram, using trigonometry, calculate the angle between the resultant and the 8.0 m displacement.

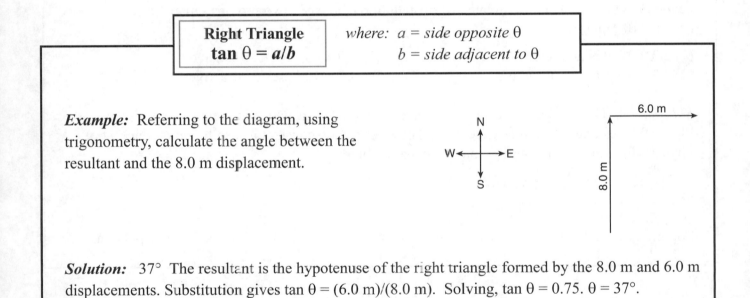

Solution: 37°   The resultant is the hypotenuse of the right triangle formed by the 8.0 m and 6.0 m displacements. Substitution gives $\tan \theta = (6.0\ \text{m})/(8.0\ \text{m})$. Solving, $\tan \theta = 0.75$. $\theta = 37°$.

Additional Information:

- These equations can be used in vector problems when asked to determine the magnitude of a component of a given vector.

Questions dealing with **Geometry and Trigonometry** will be found throughout this workbook.

Electricity deals with the interaction between charged particles, the motion of charged particles and the energy carried by moving charged particles. There are two types of charge found on particles, positive and negative.

Electrostatic force: Charged particles exert mutual forces of attraction or repulsion on each other.

The magnitude of the force between two charged particles varies directly with the charges of the two particles and inversely with the square of the distance between their centers. k is a constant called the electrostatic constant.

$$F_e = \frac{kq_1q_2}{r^2}$$

where: F_e = electrostatic force
k = electrostatic constant
q = charge
r = distance between centers

Example: What is the approximate electrostatic force between two protons separated by a distance of 1.0×10^{-6} meter?

(1) 2.3×10^{-16} N and repulsive
(2) 2.3×10^{-16} N and attractive

(3) 9.0×10^{21} N and repulsive
(4) 9.0×10^{21} N and attractive

Solution: 1 The charge on a proton is the elementary charge, which is found on the List of Physical Constants in the reference table. The electrostatic constant is also found on the List of Physical Constants. Substitution into the equation gives

$$F_e = \frac{\left(8.99 \times 10^9 \text{ N} \cdot \frac{m^2}{C^2}\right)\left(1.60 \times 10^{-19} \text{ C}\right)\left(1.60 \times 10^{-19} \text{ C}\right)}{(1.0 \times 1.0^{-6} \text{ m})^2}.$$

Solving gives 2.3×10^{-16} N. Since the protons have like charges, the force is one of repulsion.

Electrostatic forces – Additional Information:

• The value of the electrostatic constant (k) is found on the List of Physical Constants.

• The charge (q) on the particles must be expressed in coulombs (C).

• The force between like charges is one of repulsion, that between unlike charges is one of attraction.

1. Which graph best represents the electrostatic force between an alpha particle with a charge of +2 elementary charges and a positively charged nucleus as a function of their distance of separation?

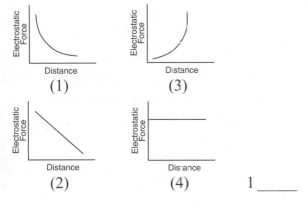

(1)

(3)

(2)

(4)   1 _____

2. The magnitude of the electrostatic force between two point charges is F. If the distance between the charges is doubled, the electrostatic force between the charges will become

(1) $\frac{F}{4}$ (3) $\frac{F}{2}$

(2) $2F$ (4) $4F$ 2 _____

3. A beam of electrons is directed into the electric field between two oppositely charged parallel plates, as shown in the diagram below.

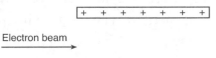

The electrostatic force exerted on the electrons by the electric field is directed

(1) into the page
(2) out of the page
(3) toward the bottom of the page
(4) toward the top of the page   3 _____

4. Two similar metal spheres, A and B, have charges of $+2.0 \times 10^{-6}$ coulomb and $+1.0 \times 10^{-6}$ coulomb, respectively, as shown in the diagram below.

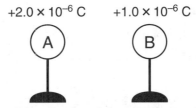

+2.0 × 10⁻⁶ C   +1.0 × 10⁻⁶ C

The magnitude of the electrostatic force on A due to B is 2.4 newtons. What is the magnitude of the electrostatic force on B due to A?

(1) 1.2 N   (3) 4.8 N
(2) 2.4 N   (4) 9.6 N   4 _____

5. A point charge of $+3.0 \times 10^{-7}$ coulomb is placed 2.0×10^{-2} meter from a second point charge of $+4.0 \times 10^{-7}$ coulomb. The magnitude of the electrostatic force between the charges is

(1) 2.7 N   (3) 3.0×10^{-10} N
(2) 5.4×10^{-2} N   (4) 6.0×10^{-12} N   5 _____

6. A distance of 1.0 meter separates the centers of two small charged spheres. The spheres exert gravitational force F_g and electrostatic force F_e on each other. If the distance between the spheres' centers is increased to 3.0 meters, the gravitational force and electrostatic force, respectively, may be represented as

(1) $\frac{F_g}{9}$ and $\frac{F_e}{9}$ (3) $3F_g$ and $3F_e$

(2) $\frac{F_g}{3}$ and $\frac{F_e}{3}$ (4) $9F_g$ and $9F_e$ 6 _____

7. What is the approximate electrostatic force between two protons separated by a distance of 1.0×10^{-6} meter?

    (1) 2.3×10^{-16} N and repulsive
    (2) 2.3×10^{-16} N and attractive
    (3) 9.0×10^{21} N and repulsive
    (4) 9.0×10^{21} N and attractive   7 _____

8. A balloon is rubbed against a student's hair and then touched to a wall. The balloon "sticks" to the wall due to

    (1) electrostatic forces between the particles of the balloon
    (2) magnetic forces between the particles of the wall
    (3) electrostatic forces between the particles of the balloon and the particles of the wall
    (4) magnetic forces between the particles of the balloon and the particles of the wall   8 _____

    Base your answer to question 9 on the information and diagram below.

    Two small metallic spheres, A and B, are separated by a distance of 4.0×10^{-1} meter, as shown. The charge on each sphere is $+1.0 \times 10^{-6}$ coulomb. Point P is located near the spheres.

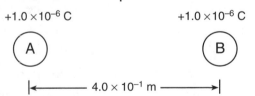

9. What is the magnitude of the electrostatic force between the two charged spheres?

    (1) 2.2×10^{-2} N   (3) 2.2×10^{4} N
    (2) 5.6×10^{-2} N   (4) 5.6×10^{4} N 9 _____

10. The diagram below shows two small metal spheres, A and B. Each sphere possesses a net charge of 4.0×10^{-6} coulomb. The spheres are separated by a distance of 1.0 meter.

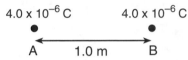

Which combination of charged spheres and separation distance produces an electrostatic force of the same magnitude as the electrostatic force between spheres A and B?

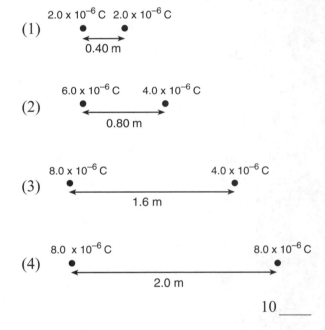

          10 _____

11. An electrostatic force of magnitude F exists between two metal spheres having identical charge q. The distance between their centers is r. Which combination of changes would produce no change in the electrostatic force between the spheres?

    (1) doubling q on one sphere while doubling r
    (2) doubling q on both spheres while doubling r
    (3) doubling q on one sphere while halving r
    (4) doubling q on both spheres while halving r

          11 _____

Base your answers to questions 12a and b on the information below.

The centers of two small charged particles are separated by a distance of 1.2×10^{-4} meter. The charges on the particles are $+8.0 \times 10^{-19}$ coulomb and $+4.8 \times 10^{-19}$ coulomb, respectively.

12. a) Calculate the magnitude of the electrostatic force between these two particles. [Show all work, including the equation and substitution with units.]

   b) Sketch a graph below showing the relationship between the magnitude of the electrostatic force between the two charged particles and the distance between the centers of the particles.

13. How do electrostatic forces differ from gravitational forces?

_____

_____

1. 1   This equation $F_e = \dfrac{kq_1q_2}{r^2}$ shows that the electrostatic force between two charges varies inversely with the square of the distance between their centers. Graph 1 shows an inverse square relationship.

2. 1   According to the equation $F_e = \dfrac{kq_1q_2}{r^2}$, the electrostatic force varies inversely with the square of the distance between the charges. Therefore if the distance doubles, the force becomes 1/4 as great.

3. 4   Electrons are negatively charges particles. Therefore, they will be repelled by the bottom plate and attracted by the top plate.

4. 2   Newton's Third Law, the Law of Action-Reaction, states that for every action, there is an equal but opposite reaction. This law holds true for all types of forces. If the force on sphere A due to sphere B is 2.4 N, the force of sphere B on sphere A must be 2.4 N.

5. 1   Under Electricity, find the equation $F_e = \dfrac{kq_1q_2}{r^2}$. The value of k is found on the List of Physical Constants. Substitution into the equation gives
$F_e = (8.99 \times 10^9 \text{ N} \bullet \text{m}^2/\text{C}^2)(3.0 \times 10^{-7} \text{ C})(4.0 \times 10^{-7} \text{ C})/(2.0 \times 10^{-2} \text{ m})^2$.
Solving, $F_e = 2.7$ N.

6. 1   Under Mechanics, find the equation $F_g = \dfrac{Gm_1m_2}{r^2}$ and under Electricity, find the equation $F_e = \dfrac{kq_1q_2}{r^2}$. Both of these equations indicate that the force involved varies inversely with the square of the distance between the centers of the objects. The distance between the centers is tripled (1.0 m to 3.0 m). The forces are then reduced to 1/9 of the original value.

Electric Field Strength: Any charged particle is surrounded by an electric field. This is a region in which another charged particle experiences a force of electrical nature acting on it.

The magnitude of the electric field strength or intensity at a point in an electric field is the magnitude of the electrostatic force acting on a charged particle at that point divided by the charge on that particle.

$$E = \frac{F_e}{q}$$

where: E = electric field strength
F_e = electrostatic force
q = charge

Example: An object with a net charge of 4.80×10^{-6} coulomb experiences an electrostatic force having a magnitude of 6.00×10^{-2} newton when placed near a negatively charged metal sphere. What is the electric field strength at this location?

(1) 1.25×10^4 N/C directed away from the sphere
(2) 1.25×10^4 N/C directed toward the sphere
(3) 2.88×10^{-8} N/C directed away from the sphere
(4) 2.88×10^{-8} N/C directed toward the sphere

Solution: 2   Electric field strength is a vector quantity, therefore possessing both magnitude and direction. Substitution into the equation gives: $E = \dfrac{(6.00 \times 10^{-2} \text{ N})}{(4.80 \times 10^{-6} \text{ C})} = 1.25 \times 10^4 \dfrac{\text{N}}{\text{C}}$.

This gives the magnitude of the electric field. By definition, the direction of an electric field shows the path taken by a free positive charge when placed in the field. The direction of an electric field is therefore from positive to negative. The field is then directed toward the negatively charged sphere.

Electric Field Strength – Additional Information:

- The charge (q) must be expressed in coulombs (C).

- An electric field is a vector quantity. This equation gives the magnitude of the electric field. By definition, the direction of an electric field is from positive to negative.

- An electric field may be represented by lines of force, which show the direction of the electric field. The field strength is proportional to the number of lines of force per unit area. These lines are perpendicular to the surface of the charged object.

- The field strength or intensity around a point charge or a charged sphere varies inversely with the square of the distance from the point or sphere.

- The electric field strength between oppositely charged parallel plates is uniform or constant.

- The electric field strength around a charged rod or cylinder varies inversely with the distance from the rod or cylinder.

1. Which graph best represents the relationship between the magnitude of the electric field strength, E, around a point charge and the distance, r, from the point charge?

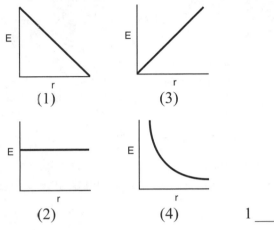

(1)   (3)

(2)   (4)   1 _____

2. Which is a vector quantity?
   (1) electric charge
   (2) electrical resistance
   (3) electrical potential difference
   (4) electric field intensity   2 _____

3. An electrostatic force of 20. newtons is exerted on a charge of 8.0×10^{-2} coulomb at point P in an electric field. The magnitude of the electric field intensity at P is

   (1) 4.0×10^{-3} N/C
   (2) 1.6 N/C
   (3) 20. N/C
   (4) 2.5×10^2 N/C   3 _____

4. The diagram below represents an electron within an electric field between two parallel plates that are charged with a potential difference of 40.0 volts.

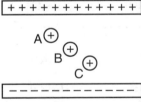

   If the magnitude of the electric force on the electron is 2.00×10^{-15} newton, the magnitude of the electric field strength between the charged plates is

   (1) 3.20×10^{-34} N/C
   (2) 2.00×10^{-14} N/C
   (3) 1.25×10^4 N/C
   (4) 2.00×10^{16} N/C   4 _____

5. Identical charges A, B, and C are located between two oppositely charged parallel plates, as shown in the diagram below.

   $+ + + + + + + + + + + +$

   $A \oplus$
   $B \oplus$
   $C \oplus$

   $- - - - - - - - - - - - -$

   The magnitude of the force exerted on the charges by the electric field between the plates is

   (1) least on A and greatest on C
   (2) greatest on A and least on C
   (3) the same on A and C, but less on B
   (4) the same for A, B, and C 5 _____

Base your answers to questions 6 a and b on the information below.

The magnitude of the electric field strength between two oppositely charged parallel metal plates is 2.0×10^3 newtons per coulomb. Point P is located midway between the plates.

6. a) On the diagram below, sketch at least five electric field lines to represent the field between the two oppositely charged plates.
   [Draw an arrowhead on each field line to show the proper direction.]

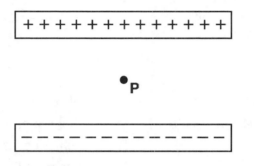

   b) An electron is located at point P between the plates. Calculate the magnitude of the force exerted on the electron by the electric field.
   [Show all work, including the equation and substitution with units.]

7. Two oppositely charged parallel metal plates, 1.00 centimeter apart, exert a force with a magnitude of 3.60×10^{-15} newton on an electron placed between the plates. Calculate the magnitude of the electric field strength between the plates. [Show all work, including the equation and substitution with units.]

8. Which type of field is present near a moving electric charge?

    (1) an electric field, only
    (2) a magnetic field, only
    (3) both an electric field and a
        magnetic field
    (4) neither an electric field nor a
        magnetic field   8 _____

10. What is the magnitude of the electrostatic force acting on an electron located in an electric field having a strength of 5.0×10^3 newtons per coulomb?

    (1) 3.1×10^{22} N
    (2) 5.0×10^3 N
    (3) 8.0×10^{-16} N
    (4) 3.2×10^{-23} N   10 _____

9. Gravitational field strength is to newtons per kilogram as electric field strength is to

    (1) coulombs per joule
    (2) coulombs per newton
    (3) joules per coulomb
    (4) newtons per coulomb   9 _____

11. What is the magnitude of the electric field intensity at a point where a proton experiences an electrostatic force of magnitude 2.30×10^{-25} newton?

    (1) 3.68×10^{-44} N/C
    (2) 1.44×10^{-6} N/C
    (3) 3.68×10^6 N/C
    (4) 1.44×10^{44} N/C   11 _____

12. The accompanying diagram represents two electrons, e_1 and e_2, located between two oppositely charged parallel plates. Compare the magnitude of the force exerted by the electric field on e_1 to the magnitude of the force exerted by the electric field on e_2.

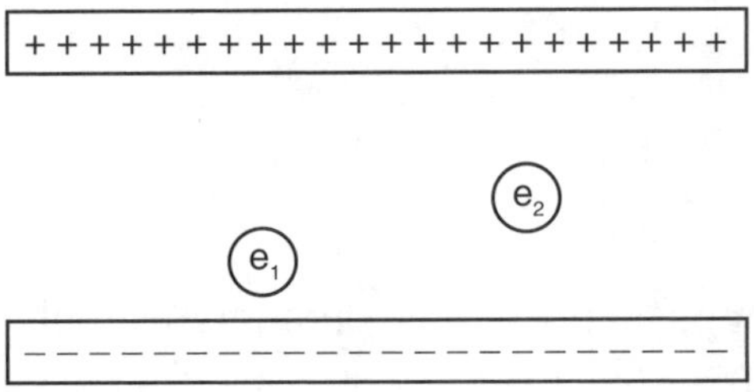

$+ +$

e_2

e_1

$- -$

Base your answers to question 13 on the information below.

The centers of two small charged particles are separated by a distance of 1.2×10^{-4} meter. The charges on the particles are $+8.0 \times 10^{-19}$ coulomb and $+4.8 \times 10^{-19}$ coulomb, respectively.

13. On the diagram below, draw at least four electric field lines in the region between the two positively charged particles.

8.0×10^{-19} C (+)   (+) 4.8×10^{-19} C

1. 4   The electric field strength around a point charge varies inversely with the square of the distance from the point charge. Graph 4 shows an inverse square relationship

2. 4   Of the choices given, only electric field strength has both magnitude and direction, making it a vector quantity.

3. 4   Under Electricity, find the equation $E = F_e/q$. Substitution gives $E = (20. \text{ N})/(8.0 \times 10^{-2} \text{ C})$. Solving, $E = 2.5 \times 10^{2}$ N/C.

4. 3   Under Electricity, find the equation $E = F_e/q$. The charge on an electron is one elementary charge, which is found in the List of Physical Constants on the reference table. Substitution and solving gives $E = (2.00 \times 10^{-15} \text{ N})/(1.60 \times 10^{-19} \text{ C}) = 1.25 \times 10^4$ N/C.

5. 4   The electric field strength is uniform between parallel charged plates. Therefore, the magnitude of the electric force acting on particles of equal charge will be the same at all points in the electric field.

6. *a)*

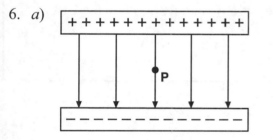

Explanation: Electric field lines are directed from the region of a positive charge toward the region of a negative charge. These lines are perpendicular to the surface of the charged objects and are therefore parallel to each other between parallel charged plates.

   b)
$$E = \frac{F_e}{q}$$
$$F_e = Eq$$
$$F_e = (2.0 \times 10^3 \text{ N/C}) \ (1.6 \times 10^{-19} \text{ C})$$
$$F_e = 3.2 \times 10^{-16} \text{ N}$$

Explanation: Under Electricity, find the equation $E = F_e/q$. The value of q is the charge on the electron in C, which is found on the List of Physical Constants.

7.
$$E = \frac{F_e}{q}$$
$$E = \frac{3.60 \times 10^{-15} \text{ N}}{1.60 \times 10^{-19} \text{ C}}$$
$$E = 2.25 \times 10^4 \text{ N/C}$$

Explanation: Under Electricity, find the equation $E = F_e/q$. The charge q is the charge on the electron which is the elementary charge (e), found on the List of Physical Constants. Substitute into the equation and solve for E.

Potential Difference: A charged particle in an electric field possesses electric potential energy. The electric potential difference or simply potential difference between two points in an electric field is the work per unit charge required to move a charge between the two points.

The potential difference between two points in an electric field is equal to the work required to move a charged particle between the two points divided by the charge on the particle.

$$V = \frac{W}{q}$$

where: V = potential difference
W = work (electrical energy)
q = charge

Example: If 4.8×10^{-17} joule of work is required to move an electron between two points in an electric field, what is the electric potential difference between these points?

(1) 1.6×10^{-19} V   (2) 4.8×10^{-17} V   (3) 3.0×10^2 V   (4) 4.8×10^2 V

Solution: 3   The charge of the electron is one elementary charge, which is found on the List of Physical Constants in the reference table. Then $V = \dfrac{4.8 \times 10^{-17} \text{ J}}{1.60 \times 10^{-19} \text{ C}} = 3.0 \times 10^2$ V.

Potential Difference – Additional Information:

- The charge (q) must be expressed in coulombs (C).

- The unit of potential difference, J/C, is defined to be the volt (V).

- The electronvolt (eV) is the energy or work required to move one elementary charge (e) between two points differing in electric potential by 1 volt.

1. If 20 joules of work is used to transfer 20 coulombs of charge through a 20-ohm resistor, the potential difference across the resistor is

   (1) 1 V   (3) 0.05 V
   (2) 20 V   (4) 400 V   1 _____

2. How much work is required to move a single electron through a potential difference of 100. volts?

   (1) 1.6×10^{-21} J   (3) 1.6×10^{-17} J
   (2) 1.6×10^{-19} J   (4) 1.0×10^2 J   2 _____

3. In an electric field, 0.90 joule of work is required to bring 0.45 coulomb of charge from point A to point B. What is the electric potential difference between points A and B?

   (1) 5.0 V   (3) 0.50 V
   (2) 2.0 V   (4) 0.41 V   3 _____

4. A potential difference of 10.0 volts exists between two points, A and B, within an electric field. What is the magnitude of charge that requires 2.0×10^{-2} joule of work to move it from A to B?

   (1) 5.0×10^2 C   (3) 5.0×10^{-2} C
   (2) 2.0×10^{-1} C   (4) 2.0×10^{-3} C   4 _____

5. If 7.68×10^{-17} joule of work is required to move an electron between two points in an electric field, what is the electric potential difference between these points?

   (1) 1.6×10^{-19}V   (3) 3.0×10^2 V
   (2) 4.8×10^{-17}V   (4) 4.8×10^2 V   5 _____

6. If a 1.5-volt cell is to be completely recharged, each electron must be supplied with a minimum energy of

   (1) 1.5 eV   (3) 9.5×10^{18} eV
   (2) 1.5 J   (4) 9.5×10^{18} J   6 _____

Base your answers to question 7 on the information below.

   A proton starts from rest and gains 8.35×10^{-14} joule of kinetic energy as it accelerates between points A and B in an electric field.

7. Calculate the potential difference between points A and B in the electric field. [Show all work, including the equation and substitution with units.]

8. The graph below shows the relationship between the work done on a charged body in an electric field and the net charge on the body.

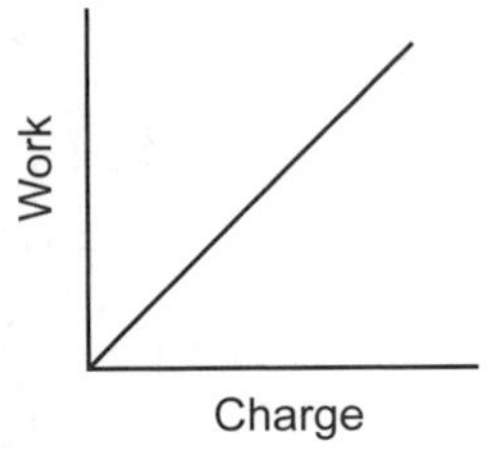

What does the slope of this graph represent?

(1) power
(2) potential difference
(3) force
(4) electric field intensity   8 _____

9. The diagram below shows proton P located at point A near a positively charged sphere.

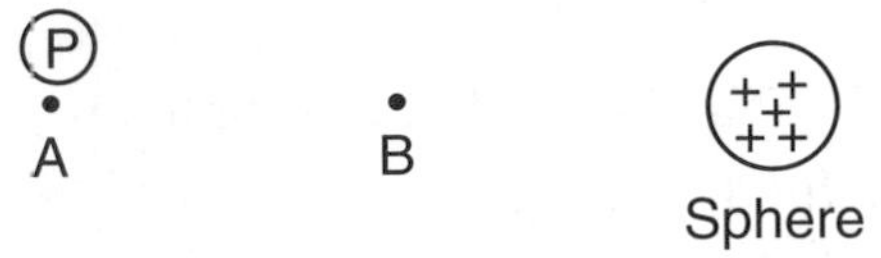

If 6.4×10^{-19} joule of work is required to move the proton from point A to point B, the potential difference between A and B is

(1) 6.4×10^{-19} V   (3) 6.4 V
(2) 4.0×10^{-19} V   (4) 4.0 V   9 _____

10. How much work is done in moving 5.0 coulombs of charge against a potential difference of 12 volts?

(1) 2.4 J   (3) 30. J
(2) 12 J   (4) 60. J   10 _____

11. If 15 joules of work is required to move 3.0 coulombs of charge between two points, the potential difference between these two points is

(1) 45 V   (3) 3.0 V
(2) 15 V   (4) 5.0 V   11 _____

12. How much electrical energy is required to move a 4.00-microcoulomb charge through a potential difference of 36.0 volts?

(1) 9.00×10^6 J
(2) 144 J
(3) 1.44×10^{-4} J
(4) 1.11×10^{-7} J   12 _____

13. An electron is accelerated through a potential difference of 2.5×10^4 volts in the cathode ray tube of a computer monitor. Calculate the work, in joules, done on the electron. [Show all work, including the equation and substitution with units.]

Copyright © 2011
Topical Review Book Company

Potential Difference
Answers
Set 1

1. 1   In this problem, assume that the resistance remains constant. Under Electricity, find the equation $V = W/q$. Substitution gives $V = \dfrac{20 \text{ J}}{20 \text{ C}}$. Solving, V = 1 V.

2. 3   Under Electricity, find the equation $V = W/q$. The charge on the electron is found in the List of Physical Constants on the reference table (the elementary charge).
Substitution gives 100. $V = \dfrac{W}{(1.60 \times 10^{-19} \text{ C})}$. Solving, W = 1.6×10^{-17} J.

3. 2   Under Electricity, find the equation $V = W/q$. Substitution and solving gives $V = \dfrac{0.90 \text{ J}}{0.45 \text{ C}} = 2.0$ V.

4. 4   Under Electricity, find the equation $V = W/q$. Substitution gives (10.0 V) = $(2.0 \times 10^{-2}$ J)/(q). Solving, q = 2.0×10^{-3} C.

5. 4   Under Electricity, find the equation $V = W/q$. The charge must be in coulombs (C). The charge of the electron is one elementary charge, which is found on the List of Physical Constants.
Then $V = \dfrac{7.68 \times 10^{-17} \text{ J}}{1.60 \times 10^{-19} \text{ C}} = 4.8 \times 10^2$ V.

6. 1   Under Electricity, find the equation $V = W/q$. Solving for W gives $W = (V)(q)$. To calculate energy in eV (electron volts), q must be in elementary charge units, which for an electron is 1 e. Substitution gives W = (1 e)(1.5 V) = 1.5 eV.

7. $V = \dfrac{W}{q}$

   $V = \dfrac{8.35 \times 10^{-14} \text{ J}}{1.60 \times 10^{-19} \text{ C}}$

   $V = 5.22 \times 10^5$ J/C   or 5.22×10^5 V

   Explanation: Work and energy are equivalent quantities. The charge on the electron is found on the List of Physical Constants. Substitution into the equation $V = W/q$ gives the setup shown.

Electricity – Electric Current

Electric Current: An electric current is a flow of charged particles from one point to another.

Quantitatively, electric current is a measure of the rate of flow of charge between two points in a conductor and equals the quantity of charge transferred between the two points divided by the time needed for the transfer.

$$I = \frac{\Delta q}{t}$$

where: I = *current*
Δq = *change in charge*
t = *time*

Example: During a thunderstorm, a lightning strike transfers 12 coulombs of charge in 2.0×10^{-3} second. What is the average current produced in this strike?

(1) 1.7×10^{-4} A (2) 2.4×10^{-2} A (3) 6.0×10^{3} A (4) 9.6×10^{3} A

Solution: 3 Current is defined as the rate of flow of electric charge. Substituting into the equation gives $I = \dfrac{12\,\text{C}}{2.0 \times 10^{-3}\,\text{s}} = 6.0 \times 10^{3}\,\text{A}$.

Electric Current – Additional Information:

- The charge (q) must be in coulombs (C) and the time must be in seconds (s).
- The unit of current (C/s) is defined to be the ampere (A).

1. If 10. coulombs of charge are transferred through an electric circuit in 5.0 seconds, then the current in the circuit is

   (1) 0.50 A   (3) 15 A

   (2) 2.0 A   (4) 50. A   1 _____

2. The current through a lightbulb is 2.0 amperes. How many coulombs of electric charge pass through the lightbulb in one minute?

   (1) 60. C   (3) 120 C

   (2) 2.0 C   (4) 240 C   2 _____

3. A charge of 5.0 coulombs moves through a circuit in 0.50 second. The current in the circuit is

   (1) 2.5 A   (3) 7.0 A

   (2) 5.0 A   (4) 10. A   3 _____

4. Charge flowing at the rate of 2.50×10^{16} elementary charges per second is equivalent to a current of

   (1) 2.50×10^{13} A

   (2) 6.25×10^{5} A

   (3) 4.00×10^{-3} A

   (4) 2.50×10^{-3} A   4 _____

5. During a thunderstorm, a lightning strike transfers 24 coulombs of charge in 2.5×10^{-3} second. What is the average current produced in this strike?

   (1) 1.7×10^{-4} A

   (2) 2.4×10^{-2} A

   (3) 6.0×10^{3} A

   (4) 9.6×10^{3} A   5 _____

6. A wire carries a current of 2.0 amperes. How many electrons pass a given point in this wire in 1.0 second?

   (1) 1.3×10^{18} (3) 1.3×10^{19}

   (2) 2.0×10^{18} (4) 2.0×10^{19} 6 _____

7. The diagram below represents a simple electric circuit

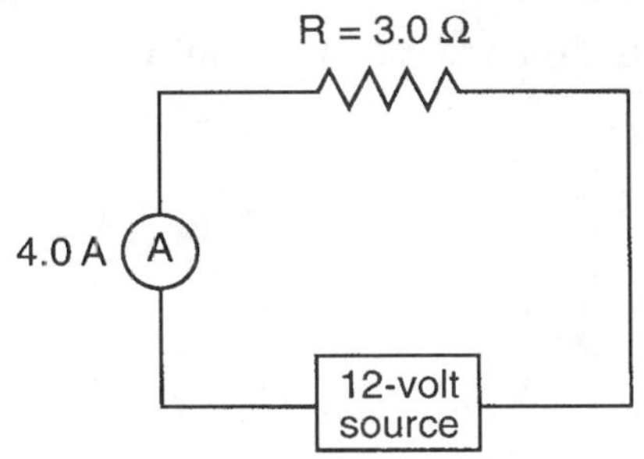

   R = 3.0 Ω

   4.0 A (A)

   12-volt source

   How much charge passes through the resistor in 2.0 seconds?

   (1) 6.0 C   (3) 8.0 C

   (2) 2.0 C   (4) 4.0 C   7 _____

8. A lightning bolt transfers 6.0 coulombs of charge from a cloud to the ground in 2.0×10^{-3} second. What is the average current during this event?

   (1) 1.2×10^{-2} A
   (2) 3.0×10^{2} A
   (3) 3.0×10^{3} A
   (4) 1.2×10^{4} A   8 _____

9. An operating lamp draws a current of 0.50 ampere. The amount of charge passing through the lamp in 10. seconds is
   (1) 0.050 C   (3) 5.0 C
   (2) 2.0 C   (4) 20. C   9 _____

10. The current traveling from the cathode to the screen in a television picture tube is 5.0×10^{-5} ampere. How many electrons strike the screen in 5.0 seconds?

    (1) 3.1×10^{24} (3) 1.6×10^{15}
    (2) 6.3×10^{18} (4) 1.0×10^{5} 10 _____

11. The accompanying diagram shows two resistors, R_1 and R_2, connected in parallel in a circuit having a 120-volt power source. Resistor R_1 develops 150 watts and resistor R_2 develops an unknown power. Ammeter A in the circuit reads 0.50 ampere. Calculate the amount of charge passing through resistor R_2 in 60. seconds. [Show all work, including the equation and substitution with units.]

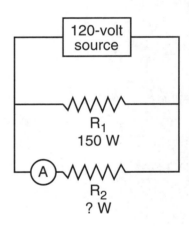

1. 2   Under Electricity, find the equation $I = \Delta q/t$. Substituting and solving gives $I = (10.\ C)/(5.0\ s) = 2.0\ A$.

2. 3   Under Electricity, find the equation $I = \Delta q/t$. In this equation, t must be in seconds. Substitution into the equation gives: $2.0\ A = \Delta q/(60.0\ s)$. Solving for Δq gives 120 C.

3. 4   Under Electricity, find the equation $I = \Delta q/t$. Substitution gives $I = \dfrac{(5.0\ C)}{(0.50\ s)}$. Solving, $I = 10.\ A$.

4. 3   Under Electricity, find the equation $I = \Delta q/t$. In this equation, Δq must be in coulombs (C). In the List of Physical Constants in the reference table, $1\ C = 6.25 \times 10^{18}$ e. Using this as a conversion factor, $2.50 \times 10^{16}\ e \times \dfrac{1\ C}{6.25 \times 10^{18}e} = 0.004\ C$. Substitution gives $I = (0.004\ C)/(1\ s)$.

   Solving, $I = 0.004\ A = 4.00 \times 10^{-3}\ A$.

5. 4   Current is defined as the rate of flow of electrical charge. Substituting into the equation gives $I = \dfrac{24\ C}{2.5 \times 10^{-3}\ s} = 9.6 \times 10^{3}\ A$.

6. 3   Use the equation $I = \Delta q/t$ to determine the amount of charge in coulombs transferred in 1.0 s. Substitution gives $2.0\ A = \Delta q/(1.0\ s)$. Solving, $\Delta q = 2.0\ C$. From the List of Physical Constants, 1 coulomb (C) $= 6.25 \times 10^{18}$ elementary charges or electrons (e^-). Use this as a conversion factor to change from C to electrons: $2.0\ C \times \dfrac{6.25 \times 10^{18}\ e^-}{1\ C} = 12.5 \times 10^{18}e^- = 1.3 \times 10^{19}e^-$.

7. 3   Under Electricity, find the equation $I = \Delta q/t$. Substitution gives $4.0\ A = \Delta q/(2.0\ s)$. Solving, $\Delta q = 8.0\ C$.

Electricity – Ohm's Law

Ohm's Law: Ohm's Law may be applied to an individual circuit element or the entire electric circuit. It relates three properties of an electric circuit: current, potential difference and resistance.

Ohm's law relates these quantities quantitatively and states the resistance is equal to potential difference divided by current.

$$R = \frac{V}{I}$$

where: R = *resistance*
V = *potential difference*
I = *current*

Example: In a simple electric circuit, a 110-volt electric heater draws 2.0 amperes of current. The resistance of the heater is

(1) 0.018 Ω (2) 28 Ω (3) 55 Ω (4) 220 Ω

Solution: 3 Direct substitution into Ohm's Law gives R = (110 V)/(2.0 A). Solving, R = 55 Ω.

Ohm's Law – Additional Information:

- The unit of electrical resistance is the ohm (Ω). From this equation, 1 Ω = 1 V/A.
- Solving for I, $I = V/R$. This indicates that the current in a circuit varies directly with the potential difference and inversely with the resistance.

1. An electric circuit contains a variable resistor connected to a source of constant potential difference. Which graph best represents the relationship between current and resistance in this circuit?

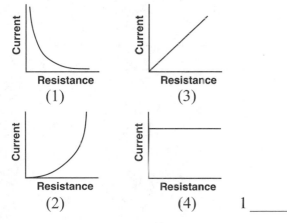

(1)   (3)

(2)   (4)   1 _____

2. A circuit consists of a resistor and a battery. Increasing the voltage of the battery while keeping the temperature of the circuit constant would result in an increase in

   (1) current, only
   (2) resistance, only
   (3) both current and resistance
   (4) neither current nor resistance   2 _____

3. A 330.-ohm resistor is connected to a 5.00-volt battery. The current through the resistor is

   (1) 0.152 mA   (3) 335 mA
   (2) 15.2 mA   (4) 1650 mA   3 _____

4. A 6.0-ohm resistor that obeys Ohm's Law is connected to a source of variable potential difference. When the applied voltage is decreased from 12 V to 6.0 V, the current passing through the resistor

   (1) remains the same
   (2) is doubled
   (3) is halved
   (4) is quadrupled   4 _____

5. What is the current in a 100.-ohm resistor connected to a 0.40-volt source of potential difference?

   (1) 250 mA   (3) 2.5 mA
   (2) 40. mA   (4) 4.0 mA   5 _____

Note: Question 6 has only three choices.

6. A student conducted an experiment to determine the resistance of a lightbulb. As she applied various potential differences to the bulb, she recorded the voltages and corresponding currents and constructed the graph below

Current vs. Potential Difference

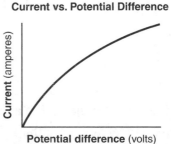

According to the graph, as the potential difference increased, the resistance of the lightbulb

   (1) decreased
   (2) increased
   (3) changed, but there is not enough information to know which way 6 _____

Base your answers to question 7 on the information and diagram below.

A 3.0-ohm resistor, an unknown resistor, R, and two ammeters, A_1 and A_2, are connected as shown with a 12-volt source. Ammeter A_2 reads a current of 5.0 amperes.

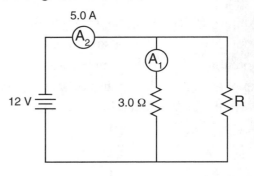

7. Determine the equivalent resistance of the circuit.

Base your answers to questions 8 *a* and *b* on the information below.

A toaster operating at 120 volts uses 8.75 amperes of current.

8. *a*) Calculate the resistance of the toaster. [Show all work, including the equation and substitution with units.]

b) The toaster is connected in a circuit protected by a 15-ampere fuse. (The fuse will shut down the circuit if it carries more than 15 amperes.) Is it possible to simultaneously operate the toaster and a microwave oven that requires a current of 10.0 amperes on this circuit? Justify your answer mathematically.

9. The graph below represents the relationship between the potential difference (V) across a resistor and the current (I) through the resistor.

    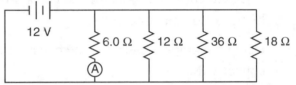

    Through which entire interval does the resistor obey Ohm's law?

    (1) *AB*   (3) *CD*
    (2) *BC*   (4) *AD*   9 _____

10. How much current flows through a 12-ohm flashlight bulb operating at 3.0 volts?

    (1) 0.25 A   (3) 3.0 A
    (2) 0.75 A   (4) 4.0 A   10 _____

11. Which graph best represents the relationship between the electrical power and the current in a resistor that obeys Ohm's Law?

    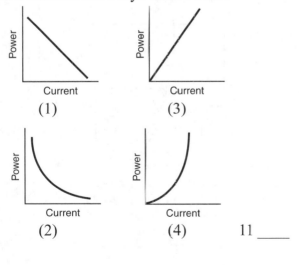

    11 _____

12. In a flashlight, a battery provides a total of 3.0 volts to a bulb. If the flashlight bulb has an operating resistance of 5.0 ohms, the current through the bulb is

    (1) 0.30 A   (3) 1.5 A
    (2) 0.60 A   (4) 1.7 A   12 _____

    Base your answers to question 13 on the diagram below, which represents an electric circuit consisting of four resistors and a 12-volt battery.

    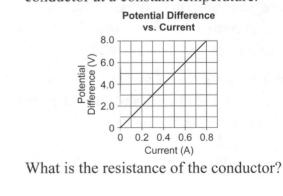

13. What is the current measured by ammeter *A*?

    (1) 0.50 A   (3) 72 A
    (2) 2.0 A   (4) 4.0 A   13 _____

14. The graph below represents the relationship between the potential difference across a metal conductor and the current through the conductor at a constant temperature.

    Potential Difference vs. Current

    What is the resistance of the conductor?

    14 _____

15. What is the current in a 100.-ohm resistor connected to a 0.40-volt source of potential difference? [Show all work, including the equation and substitution with units.]

16. A long copper wire was connected to a voltage source. The voltage was varied and the current through the wire measured, while temperature was held constant. The collected data are represented by the accompanying graph.

    Using the graph, determine the resistance of the copper wire.

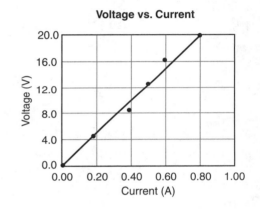

17. The accompanying diagram represents a simple circuit consisting of a variable resistor, a battery, an ammeter, and a voltmeter.

    What is the effect of increasing the resistance of the variable resistor from 1000 Ω to 10000 Ω? [Assume constant temperature.]

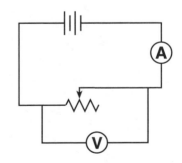

Ohm's Law

Answers

Set 1

1. 1   Under Electricity, find the equation $R = V/I$. Solving for I, $I = V/R$. If the potential difference is constant, the current varies inversely with the resistance. Graph 1 shows an inverse relationship.

2. 1   Under Electricity find the equation $R = V/I$. Rearranging gives $I = V/R$. If the temperature remains constant, the resistance remains constant. Since this shows a direct relationship, as the voltage increases, the current will increase.

3. 2   Under Electricity, find the equation $R = V/I$. Substituting into the equation gives 330. $\Omega = (5.00 \text{ V})/I$. Solving, $I = 0.0152$ A. Using the Prefixes for Powers of 10 chart in the reference table, the prefix milli (m) means 10^{-3}. Using this as a conversion factor: $0.0152 \text{ A} \times \dfrac{1\,\text{mA}}{10^{-3}\,\text{A}} = 15.2\,\text{mA}$.

4. 3   Using Ohm's Law $R = V/I$, solving for I gives $I = V/R$. From this equation, the current through a resistor varies directly with the applied voltage. Therefore, as the voltage is halved (12 V to 6.0 V), the current through the resistor is halved.

5. 4   Find the equation $R = V/I$. Substitution gives 100. $\Omega = (0.40 \text{ V})/(I)$. Solving, $I = 4.0 \times 10^{-3}$ A. From the table for Prefixes for Powers of 10 in the reference tables, the prefix for 10^{-3} is milli (m). Therefore, 4.0×10^{-3} A $= 4.0$ mA.

6. 2   The slope of the graph is $\Delta I / \Delta V$. From the reference table under Electricity, $R = V/I$. The slope of the graph is the reciprocal of the resistance $(1/R)$. Since the slope of the graph is decreasing, the resistance then must be increasing.

7. Answer: 2.4 Ω

   Under Electricity find the equation $R = V/I$ (Ohm's Law). The equivalent resistance of the circuit can be calculated using the total circuit voltage (12 V) and total circuit current (5.0 A). Substitution gives $R = \dfrac{(12 \text{ V})}{(5.0 \text{ A})}$ and $R = 2.4\ \Omega$.

8. a) Answer 13.7 Ω

   Under Electricity, find the equation $R = V/I$. Substituting and solving gives $R = (120 \text{ V})/(8.75 \text{ A}) = 13.7\ \Omega$.

   b) No

   The toaster draws 8.75 amperes of current. When this current is added to the current drawn by the microwave oven, the total current is greater than the rating of the fuse (10.0 A + 8.75 A = 18.75 A).

Electrical Resistance: This is the opposition a material offers to a flow of charge (current) through it.

The resistance of a wire varies directly with the length of the wire and inversely with the cross sectional area of the wire. ρ is a constant called resistivity and depends upon the composition of the wire.

$$R = \frac{\rho L}{A}$$

where: R = resistance
ρ = resistivity
L = length of conductor
A = cross-sectional area

Example: What is the resistance at 20°C of a 1.50-meter long aluminum conductor that has a cross-sectional area of 1.13×10^{-6} meter^2?

(1) 1.87×10^{-3} Ω   (2) 2.28×10^{-2} Ω   (3) 3.74×10^{-2} Ω   (4) 1.33×10^{6} Ω

Solution: 3   The resistivity of Al is given on Resistivities at 20°C table. Substitution into the equation gives $R = (2.82 \times 10^{-8}$ Ω•m$)(1.50$ m$)/(1.13 \times 10^{-6}$ m$^2) = 3.74 \times 10^{-2}$ Ω.

Electrical Resistance – Additional Information:

- Resistivity values (ρ) are found on the Resistivities at 20°C table.

- The length must be in meters (m) and the cross-sectional area must be in m^2.

- At temperatures near absolute zero, the resistance of materials approaches zero. This is superconductivity.

1. Several pieces of copper wire, all having the same length but different diameters, are kept at room temperature. Which graph best represents the resistance, R, of the wires as a function of their cross-sectional areas, A?

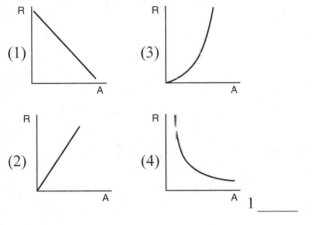

(1)

(3)

(2)

(4)

1 _____

2. What is the resistance at 20.°C of a 2.0-meter length of tungsten wire with a cross-sectional area of 7.9×10^{-7} meter^2?

(1) 5.7×10^{-1} Ω
(2) 1.4×10^{-1} Ω
(3) 7.1×10^{-2} Ω
(4) 4.0×10^{-2} Ω

2 _____

3. A copper wire of length L and cross-sectional area A has resistance R. A second copper wire at the same temperature has a length of $2L$ and a cross-sectional area of A. What is the resistance of the second copper wire?

(1) R (2) $2R$ (3) R (4) $4R$ 3 _____

4. A 12.0-meter length of copper wire has a resistance of 1.50 ohms. How long must an aluminum wire with the same cross-sectional area be to have the same resistance?

(1) 7.32 m   (3) 12.0 m
(2) 8.00 m   (4) 19.7 m   4 _____

5. A 0.500-meter length of wire with a cross-sectional area of 3.14×10^{-6} meters squared is found to have a resistance of 2.53×10^{-3} ohms. According to the resistivity chart, the wire could be made of

(1) aluminum   (3) nichrome
(2) copper   (4) silver   5 _____

6. The table below lists various characteristics of two metallic wires, A and B.

Wire	Material	Temperature (°C)	Length (m)	Cross-Sectional Area (m²)	Resistance (Ω)
A	silver	20.	0.10	0.010	R
B	silver	20.	0.20	0.020	???

If wire A has resistance R, then wire B has resistance

(1) R (2) $2R$ (3) $\frac{R}{2}$ (4) $4R$ 6 _____

7. A length of copper wire and a 1.00-meter-long silver wire have the same cross-sectional area and resistance at 20°C. Calculate the length of the copper wire. [Show all work, including the equation and substitution with units.]

Base your answers to question 8a and b on the information and diagram below.

A 10.0-meter length of copper wire is at 20°C. The radius of the wire is 1.0×10^{-3} meter.

Cross Section of Copper Wire

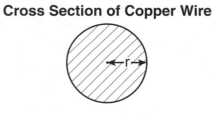

$r = 1.0 \times 10^{-3}$ m

8. *a*) Determine the cross-sectional area of the wire.

b) Calculate the resistance of the wire. [Show all work, including the equation and substitution with units.]

9. Plastic insulation surrounds a wire having diameter d and length ℓ as shown below.

   A decrease in the resistance of the wire would be produced by an increase in the

   (1) thickness of the plastic insulation
   (2) length ℓ of the wire
   (3) diameter d of the wire
   (4) temperature of the wire   9 _____

10. Which changes would cause the greatest increase in the rate of flow of charge through a conducting wire?

   (1) increasing the applied potential difference and decreasing the length of wire
   (2) increasing the applied potential difference and increasing the length of wire
   (3) decreasing the applied potential difference and decreasing the length of wire
   (4) decreasing the applied potential difference and increasing the length of wire   10 _____

Base your answers to questions 11 and 12 on the information below.

   A copper wire at 20°C has a length of 10.0 meters and a cross-sectional area of 1.00×10^{-3} meter^2. The wire is stretched, becomes longer and thinner, and returns to 20°C.

11. As a result of this stretching, the resistance of the wire will

   (1) decrease
   (2) increase
   (3) remains the same   11 _____

12. As a result of this stretching, the resistivity of the wire will

   (1) decrease
   (2) increase
   (3) remains the same   12 _____

13. A 0.686-meter-long wire has a cross-sectional area of 8.23×10^{-6} meter^2 and a resistance of 0.125 ohm at 20° Celsius. This wire could be made of

   (1) aluminum   (3) nichrome
   (2) copper   (4) tungsten   13 _____

Base your answers to question 14 a and b on the information below.

   A 1.00-meter length of nichrome wire with a cross-sectional area of 7.85×10^{-7} meter^2 is connected to a 1.50-volt battery.

14. a) Calculate the resistance of the wire. [Show all work, including the equation and substitution with units.]

b) Determine the current in the wire.

Electrical Resistance
Answers
Set 1

1. 4   Under Electricity, find the equation $R = \rho L/A$. This equation indicates that the resistance of a wire varies inversely with the cross-sectional area. Graph 4 shows an inverse relationship.

2. 2   Under Electricity, find the equation $R = \rho L/A$. The resistivity of tungsten is found in the reference table on the Resitivities at 20° C table. Substituting into the equation and solving gives $R = \dfrac{(5.6 \times 10^{-8} \, \Omega \bullet m)(2.0 \text{ m})}{(7.9 \times 10^{-7} m^2)}$.

3. 4   The equation, $R = \rho L/A$ tells us that the resistance of a wire varies directly with the length, L. The second wire is twice as long and then has twice the resistance. The equation also tells us that the resistance of a wire varies inversely with the cross-sectional area, A. The second wire has half the cross-sectional area and then has twice the resistance. The total change in resistance produced by doubling the length and halving the cross-sectional area is then $2 \times 2 = 4$. The resistance of the second wire is $4\,R$.

4. 1   Under Electricity, find the equation $R = \rho L/A$. Since the resistance and cross-section area of the two wires must be the same, we can write $(\rho L)_{Cu} = (\rho L)_{Al}$. The resistivities of Cu and Al are given on the table of Resistivities at 20° C on the reference table. Substitution gives $(1.72 \times 10^{-8} \, \Omega \bullet m)(12.0 \text{ m}) = (2.82 \times 10^{-8} \, \Omega \bullet m)(L)$. Solving, L = 7.32 m.

5. 4   Under Electricity, find the equation $R = \rho L/A$. Use this equation to solve for the resistivity of the material which is a property of the material at a given temperature and can be used to identify that material. Substitution gives $(2.53 \times 10^{-3} \, \Omega) = (\rho)(0.500 \text{ m})/(3.14 \times 10^{-6} m^2)$. Solving $\rho = 1.59 \times 10^{-8} \, \Omega \bullet m$. According to the Resistivities at 20° C table, the material is silver.

6. 1   Under Electricity, find the equation $R = \rho L/A$. Since both wires are silver at the same temperature, the resistivity is the same for both. From the equation, resistance varies directly with length and inversely with cross-sectional area. Doubling the length doubles the resistance while doubling the cross-sectional area halves the resistance. The effects cancel one another $(2 \times \frac{1}{2} = 1)$ and the resistance remains R.

7. $R = \left(\dfrac{\rho L}{A}\right)_{copper} = \left(\dfrac{\rho L}{A}\right)_{silver}$

$R = \dfrac{\rho_{copper}\, L_{copper}}{A} = \dfrac{\rho_{silver}\, L_{silver}}{A}$

$L_{copper} = \dfrac{\rho_{silver}\, L_{silver}}{\rho_{copper}}$

$L_{copper} = \dfrac{\left((1.59 \times 10^{-8}\ \Omega \bullet m)(1.00\ m)\right)}{1.72 \times 10^{-8}\ \Omega \bullet m}$

$L_{copper} = 0.924\ m$

Explanation: Using the equation $R = \rho L/A$ and the fact that the resistance of the copper wire and silver wire are the same, the first solution statement may be written. Also, the cross-section area of the two wires are the same. Therefore, they cancel from the equation. The resistivities are found on the Resistivities at 20° C table on the reference table. Substitution of the given values into the equation and solving gives L = 0.924 m.

8. *a*) Answer: $3.1 \times 10^{-6}\ m^2$

Explanation: In the reference table under Geometry and Trigonometry, find the equation $A = \pi r^2$. Substitution gives A =(3.14)(1.0 × 10^{-3} m)^2. Solving, A = 3.1 × 10^{-6} m^2.

b) $R = \dfrac{\rho L}{A}$

$R = \dfrac{(1.72 \times 10^{-8}\ \Omega \bullet m)(10.0\,m)}{3.1 \times 10^{-6}\ m^2}$

$R = 5.5 \times 10^{-2}\ \Omega$

Explanation: Under Electricity, find the equation $R = \rho L/A$. The value of ρ is found on the table of Resistivities at 20° C in the reference table. Use the value for A determined in question 8*a*.

Electrical Work, Energy and Power: Work and energy are equivalent quantities. Power is the rate at which work is done or energy is expended.

From the definition of potential difference, work done in an electric circuit can be expressed as $W = Vq$. From the definition of electric current, $q = It$. Substitution into the first equation gives $W = VIt$. Work is therefore the product of potential difference, current and time. The other parts of the equation are obtained by substitution from Ohm' Law and the definition of power.

$$W = Pt = VIt = I^2Rt = \frac{V^2t}{R}$$

where: W = work (electrical energy)
P = electrical power
t = time
V = potential difference
I = current
R = resistance

Example: An operating electric heater draws a current of 10. amperes and has a resistance of 12 ohms. How much energy does the heater use in 60. seconds?

(1) 120 J (2) 1,200 J (3) 7,200 J (4) 72,000 J

Solution: 4 Since work and energy are equivalent quantities, we can use $W = E = I^2Rt$. Substitution and solving gives $E = (10.\ \text{A})^2(12\ \Omega)(60.\ \text{s}) = 72,000$ J.

Power equals work divided by time. Dividing each of the equations given for electric work by time gives the equations for electric power. Power is a product of the potential difference across a circuit and the current in the circuit.

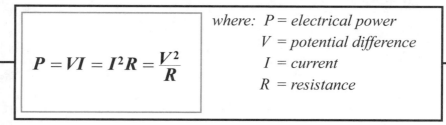

$$P = VI = I^2R = \frac{V^2}{R}$$

where: P = electrical power
V = potential difference
I = current
R = resistance

Example: An electric motor draws 150 amperes of current while operating at 240 volts. What is the power rating of this motor?

(1) 1.6 W (2) 3.8×10^2 W (3) 3.6×10^4 W (4) 5.4×10^6 W

Solution: 3 Direct substitution into the equation $P = VI$ gives $P = (240\ \text{V})(150\ \text{A})$. Solving, $P = 3.6 \times 10^4$ W.

Electrical Work, Energy and Power – Additional Information:

- The unit of electrical work is the joule (J) and that of power is the watt (W).

- The rating of light bulbs is the rate at which electrical energy is consumed or power (watts).

- Electric bills are based on the electrical energy consumed in units of kilowatt-hours (kW-h).

- In all cases, some electrical energy is converted to heat due to the inherent resistance of a circuit.

Set 1 — Electrical Work, Energy and Power

1. The potential difference applied to a circuit element remains constant as the resistance of the element is varied. Which graph best represents the relationship between power (P) and resistance (R) of this element?

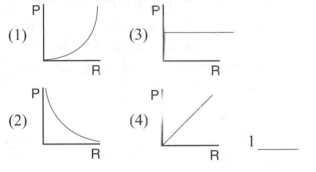

Note: Question 2 has only three choices.

2. As the potential difference across a given resistor is increased, the power expended in moving charge through the resistor

    (1) decreases
    (2) increases
    (3) remains the same   2 _____

3. One watt is equivalent to one

    (1) N•m   (3) J•s
    (2) N/m   (4) J/s   3 _____

4. The resistance of a 60.-watt lightbulb operated at 120 volts is approximately

    (1) 720 Ω   (3) 120 Ω
    (2) 240 Ω   (4) 60. Ω   4 _____

5. An immersion heater has a resistance of 5.0 ohms while drawing a current of 3.0 amperes. How much electrical energy is delivered to the heater during 200. seconds of operation?

    (1) 3.0×10^3 J   (3) 9.0×10^3 J
    (2) 6.0×10^3 J   (4) 1.5×10^4 J   5 _____

6. A device operating at a potential difference of 1.5 volts draws a current of 0.20 ampere. How much energy is used by the device in 60. seconds?

    (1) 4.5 J   (3) 12 J
    (2) 8.0 J   (4) 18 J   6 _____

7. A 4.50-volt personal stereo uses 1950 joules of electrical energy in one hour. What is the electrical resistance of the personal stereo?

    (1) 433 Ω   (3) 37.4 Ω
    (2) 96.3 Ω   (4) 0.623 Ω   7 _____

8. An operating 100.-watt lamp is connected to a 120-volt outlet. What is the total electrical energy used by the lamp in 60. seconds?

    (1) 0.60 J   (3) 6.0×10^3 J
    (2) 1.7 J   (4) 7.2×10^3 J   8 _____

9. The diagram below represents an electric circuit.

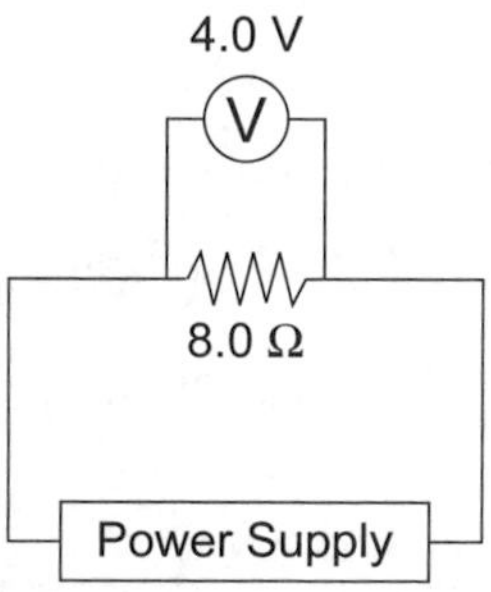

4.0 V

8.0 Ω

Power Supply

The total amount of energy delivered to the resistor in 10. seconds is

(1) 3.2 J   (3) 20. J
(2) 5.0 J   (4) 320 J   9 _____

10. What is the total electrical energy used by a 1500-watt hair dryer operating for 6.0 minutes?

(1) 4.2 J   (3) 9.0×10^3 J
(2) 250 J   (4) 5.4×10^5 J   10 _____

11. To increase the brightness of a desk lamp, a student replaces a 60-watt light bulb with a 100-watt bulb. Compared to the 60-watt bulb, the 100-watt bulb has

(1) less resistance and draws more current
(2) less resistance and draws less current
(3) more resistance and draws more current
(4) more resistance and draws less current

                                   11 _____

Note: Question 12 has only three choices.

12. An electric circuit contains a variable resistor connected to a source of constant voltage. As the resistance of the variable resistor is increased, the power dissipated in the circuit

(1) decreases
(2) increases
(3) remains the same   12 _____

13. Calculate the resistance of a 900.-watt toaster operating at 120 volts. [Show all work, including the equation and substitution with units.]

14. A light bulb attached to a 120.-volt source of potential difference draws a current of 1.25 amperes for 35.0 seconds. Calculate how much electrical energy is used by the bulb. [Show all work, including the equation and substitution with units.]

15. If the potential difference applied to a fixed resistance is doubled, the power dissipated by that resistance

    (1) remains the same (3) halves
    (2) doubles   (4) quadruples 15 _____

16. How much time is required for an operating 100-watt light bulb to dissipate 10 joules of electrical energy?

    (1) 1 s   (3) 10 s
    (2) 0.1 s   (4) 1000 s   16 _____

17. An electric drill operating at 120. volts draws a current of 3.00 amperes. What is the total amount of electrical energy used by the drill during 1.00 minute of operation?

    (1) 2.16×10^4 J   (3) 3.60×10^2 J
    (2) 2.40×10^3 J   (4) 4.00×10^1 J   17 _____

18. An electric iron operating at 120 volts draws 10. amperes of current. How much heat energy is delivered by the iron in 30. seconds?

    (1) 3.0×10^2 J   (3) 3.6×10^3 J
    (2) 1.2×10^3 J   (4) 3.6×10^4 J   18 _____

19. A potential drop of 50. volts is measured across a 250-ohm resistor. What is the power developed in the resistor?

    (1) 0.20 W   (3) 10. W
    (2) 5.0 W   (4) 50. W   19 _____

20. A 50-watt lightbulb and a 100-watt lighbulb are each operated at 110 volts. Compared to the resistance of the 50-watt bulb, the resistance of the 100-watt bulb is

    (1) half as great
    (2) twice as great
    (3) one-fourth as great
    (4) four times as great   20 _____

21. A generator produces a 115-volt potential difference and a maximum of 20.0 amperes of current. Calculate the total electrical energy the generator produces operating at maximum capacity for 60. seconds. [Show all work, including the equation and substitution with units.]

Base your answers to questions 22 *a* through *e* on the information and data table below.

A variable resistor was connected to a battery. As the resistance was adjusted, the current and power in the circuit were determined. The data are recorded in the accompanying table.

Current (amperes)	Power (watts)
0.75	2.27
1.25	3.72
2.25	6.75
3.00	9.05
4.00	11.9

Using the information in the data table, construct a line graph on the grid below, following the directions below.

Power vs. Current for a Variable Resistor

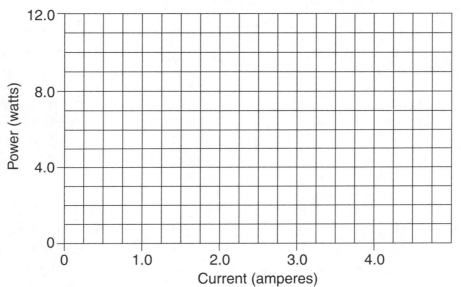

22. *a*) Plot the data points for power versus current.

b) Draw the best-fit line.

c) Using your graph, determine the power delivered to the circuit at a current of 3.5 amperes. _____ W

d) Calculate the slope of the graph. [Show all calculations, including the equation and substitution with units.]

e) What is the physical significance of the slope of the graph? _____

_____

Electrical Work, Energy and Power
Answers
Set 1

1. **2** Under Electricity, find the equation $P = V^2/R$. This equation shows that at a constant potential difference, the power varies inversely with the resistance. Graph 2 shows an inverse relationship.

2. **2** Under Electricity, find the equation $P = V^2/R$. If R remains constant, the power varies directly with the square of the potential difference. Therefore, as the potential difference increases, the power expended increases.

3. **4** The watt is a unit of power. Under Electricity, find the equation $W = Pt$. Solving for P, $P = W/t$. The unit for work is the J and that for time is the s. Therefore, the watt is a J/s.

4. **2** Substituting into $P = V^2/R$, gives 60. W = $(120. V)^2/R$. Solving for R yields 240 Ω.

5. **3** Under Electricity, find the equation $W = I^2Rt$. Work and energy are equivalent quantities in this equation. Substitution into the equation gives $W = (3.0 A)^2(5.0 \Omega)(200. s)$. Solving, $W = 9.0 \times 10^3$ J.

6. **4** Energy and work are equivalent quantities. Under Electricity in the reference table, find the equation $W = VIt$. Substitution gives $W = (1.5 V)(0.20 A)(60. s) = 18$ J.

7. **3** Work and energy are equivalent quantities. In the reference table under Electricity, find the equation $W = V^2t/R$. The time must be in seconds (1h = 3600 s). Substitution gives 1950 J = $(4.50 V)^2(3600 s)/R$. Solving, $R = 37.4 \Omega$.

8. **3** Under Electricity, find the equation $W = Pt$. Substitution gives $W = (100. W)(60. s)$. Solving, $W = 6.0 \times 10^3$ J.

9. **3** Work and energy are equivalent quantities. Under Electricity, find the equation $W = V^2t/R$. Substitution gives $W = E = (4.0 V)^2(10. s)/(8.0 \Omega)$. Solving, $E = 20.$ J.

10. **4** Under Electricity, find the equation $W = Pt$. In this equation, time must be in seconds (6.0 min = 360 s). Substitution gives $W = (15 W)(360 s)$. Solving, $W = 5.4 \times 10^4$.

11. **1** Under Electricity, find the equations $P = VI$ and $W = V^2t/R$. With a constant potential difference, the first equation shows that power varies directly with the current and the second equation shows that power varies inversely with the resistance. Therefore, a higher wattage bulb will have less resistance and draw a greater current.

12. 1 The equation $P = V^2/R$. This indicates that the power dissipated varies inversely with the resistance. Therefore, as the resistance increases, the power decreases.

13. Answer: 16 Ω

In the reference table under Electricity, find the equation $P = V^2/R$. Substitution gives $900\ W = (120\ V)^2/(R)$. Solving, $R = 16\ \Omega$.

14. $W = VIt$
$W = (120.\ \text{V})(1.25\ \text{A})(35.0\ \text{s})$
$W = 5250\ \text{J}$

or

$R = \dfrac{V}{I}$

$R = \dfrac{120.\ \text{V}}{1.25\ \text{A}} = 96.0\ \Omega$

$E = I^2Rt = (1.25\ \text{A})^2(96.0\ \Omega)(35.0\ \text{s}) = 5250\ \text{J}$

or

$R = \dfrac{V}{I}$

$R = \dfrac{120.\ \text{V}}{1.25\ \text{A}} = 96.0\ \Omega$

$E = \dfrac{V^2 t}{R} = \dfrac{(120.\ \text{V})^2(35.0\ \text{s})}{96.0\ \Omega} = 5250\ \text{J}$

Explanation: Under Electricity, find the equation $W = VIt$. Since work and energy are equivalent quantities, the equation may be written as $E = VIt$. Substitution of the values into the equation gives the energy in J.

or

From the same table, use $R = V/I$ to solve for the resistance R. Then use the equations $E = I^2Rt$ or $E = V^2t/R$ to find energy.

Series Circuits: These circuits provide a single path for current.

Since there is only a single path for current, the current through each component is the same as the total circuit current (I).

$$I = I_1 = I_2 = I_3 = \dots \quad \text{where: } I = current$$

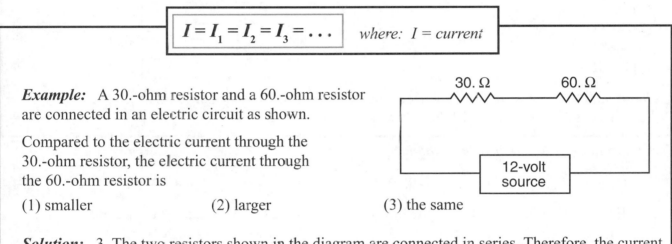

Example: A 30.-ohm resistor and a 60.-ohm resistor are connected in an electric circuit as shown.

Compared to the electric current through the 30.-ohm resistor, the electric current through the 60.-ohm resistor is

(1) smaller   (2) larger   (3) the same

Solution: 3 The two resistors shown in the diagram are connected in series. Therefore, the current through each resistor is the same.

The sum of the potential differences across each component is equal to the total potential difference applied to the circuit (V).

$$V = V_1 + V_2 + V_3 + \dots \quad \text{where: } V = potential\ difference$$

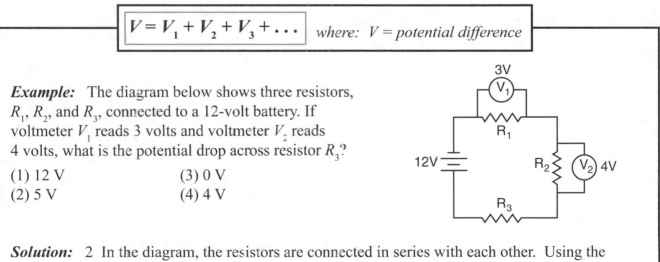

Example: The diagram below shows three resistors, R_1, R_2, and R_3, connected to a 12-volt battery. If voltmeter V_1 reads 3 volts and voltmeter V_2 reads 4 volts, what is the potential drop across resistor R_3?

(1) 12 V   (3) 0 V
(2) 5 V   (4) 4 V

Solution: 2 In the diagram, the resistors are connected in series with each other. Using the equation for total potential difference in a series circuit, $12\,V = 3\,V + 4\,V + V_3$. Solving, $V_3 = 5\,V$.

The total or equivalent resistance of resistors in series is equal to the sum of the individual resistances.

$$R_{eq} = R_1 + R_2 + R_3 + \ldots$$

where: R_{eq} = equivalent resistance
R = resistance

Example: What is the equivalent resistance shown in the accompanying diagram?

(1) 30. Ω
(2) 20. Ω
(3) 10. Ω
(4) 2.0 Ω

Solution: 1   The resistors are connected in series. As shown in the above equation, the equivalent resistance is R_{eq} = (20.Ω) + (10.Ω) = 30.Ω

Series Circuits – Additional Information:

- $I = I_1 = I_2 = I_3 = \ldots$ This is the Law of Conservation of Charge applied to a series circuit.

- $V = V_1 + V_2 + V_3 + \ldots$ This is the Law of Conservation of Energy applied to a series circuit.

- From the equation for R_{eq} in series (a sum relationship), the equivalent resistance of resistors in series will be greater than the largest resistance.

Set 1 — Series Circuits

1. Which combination of resistors has the smallest equivalent resistance?

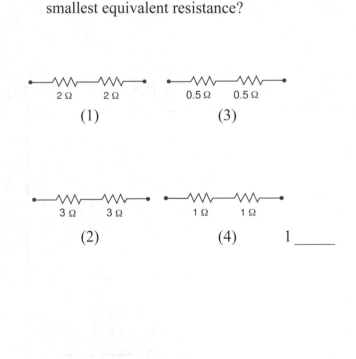

(1) 2 Ω   2 Ω

(3) 0.5 Ω   0.5 Ω

(2) 3 Ω   3 Ω

(4) 1 Ω   1 Ω   1 _____

2. Three identical lamps are connected in series with each other. If the resistance of each lamp is X ohms, what is the equivalent resistance of this series combination?

(1) X Ω

(2) $\frac{X}{3}$ Ω

(3) 3X Ω

(4) $\frac{3}{X}$ Ω   2 _____

3. Three resistors, 4 ohms, 6 ohms, and 8 ohms, are connected in series in an electric circuit. The equivalent resistance of the circuit is

(1) less than 4 Ω
(2) between 4 Ω and 6 Ω
(3) between 6 Ω and 18 Ω
(4) greater 8 Ω   3 _____

4. The diagram below shows a circuit with two resistors.

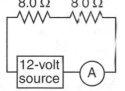

8.0 Ω   8.0 Ω

12-volt source   A

What is the reading on ammeter *A*?

(1) 1.3 A   (3) 3.0 A
(2) 1.5 A   (4) 0.75 A   4 _____

6. In the circuit shown below, voltmeter V_2 reads 80. volts.

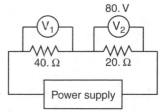

80. V

V_1 V_2

40. Ω   20. Ω

Power supply

What is the reading of voltmeter V_1?

(1) 160 V   (3) 40. V
(2) 80. V   (4) 20. V   6 _____

5. The diagram below represents an electric circuit consisting of a 12-volt battery, a 3.0-ohm resistor, R_1, and a variable resistor, R_2.

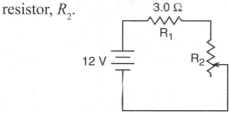

3.0 Ω
R_1

12 V R_2

At what value must the variable resistor be set to produce a current of 1.0 ampere through R_1?

(1) 6.0 Ω   (3) 3.0 Ω
(2) 9.0 Ω   (4) 12 Ω   5 _____

7. As the number of resistors in a series circuit is increased, what happens to the equivalent resistance of the circuit and total current in the circuit?

(1) Both equivalent resistance and total current decrease.
(2) Both equivalent resistance and total current increase.
(3) Equivalent resistance decreases and total current increases.
(4) Equivalent resistance increases and total current decreases.   7 _____

8. The accompanying diagram represents an electric circuit. Calculate the total amount of energy delivered to the circuit in 10. seconds. [Show all work, include the equation, and substitution]

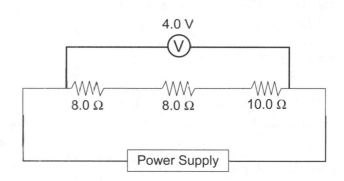

4.0 V
V

8.0 Ω   8.0 Ω   10.0 Ω

Power Supply

9. The diagram below shows three resistors, R_1, R_2, and R_3, connected to a 24-volt battery.

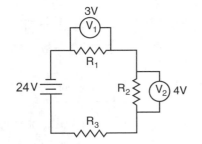

   If voltmeter V_1 reads 3 volts and voltmeter V_2 reads 4 volts, what is the potential drop across resistor R_3?

   (1) 12 V   (3) 0 V
   (2) 17 V   (4) 4 V   9 _____

10. A 10.-ohm resistor and a 20.-ohm resistor are connected in series to a voltage source. When the current through the 10.-ohm resistor is 2.0 amperes, what is the current through the 20.-ohm resistor?

   (1) 1.0 A   (3) 0.50 A
   (2) 2.0 A   (4) 4.0 A   10 _____

11. The diagram below shows two resistors connected in series to a 20.-volt battery.

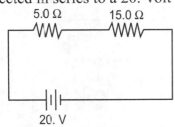

   If the current through the 5.0-ohm resistor is 1.0 ampere, the current through the 15.0-ohm resistor is

   (1) 1.0 A   (3) 3.0 A
   (2) 0.33 A   (4) 1.3 A   11 _____

12. A 9.0-volt battery is connected to a 4.0-ohm resistor and a 5.0-ohm resistor as shown in the diagram below.

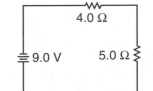

   What is the current in the 5.0-ohm resistor?

   (1) 1.0 A   (3) 2.3 A
   (2) 1.8 A   (4) 4.0 A   12 _____

13. The diagram below shows a circuit with three resistors.

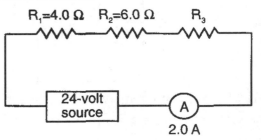

   What is the resistance of resistor R_3?

   (1) 6.0 R   (3) 12 R
   (2) 2.0 R   (4) 4.0 R   13 _____

14. A 3.0-ohm resistor and a 6.0-ohm resistor are connected in series in an operating electric circuit. If the current through the 3.0-ohm resistor is 4.0 amperes, what is the potential difference across the 6.0-ohm resistor?

   (1) 8.0 V   (3) 12 V
   (2) 2.0 V   (4) 24 V   14 _____

Base your answers to question 15a through c on the information and diagram below.

A 50.-ohm resistor, an unknown resistor R, a 120-volt source, and an ammeter are connected in a complete circuit. The ammeter reads 0.50 ampere.

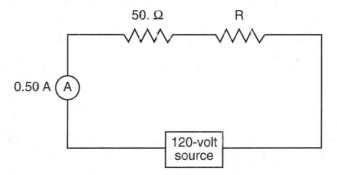

15. *a*) Calculate the equivalent resistance of the circuit. [Show all work, including the equation and substitution with units.]

b) Determine the resistance of resistor R.

c) Calculate the power dissipated by the 50.-ohm resistor. [Show all work, including the equation and substitution with units.]

Electricity
Page 149

Series Circuits
Answers
Set 1

1. 3 For series circuits, $R_{eq} = R_1 + R_2 + R_3 + \ldots$. For choice 1, $R_{eq} = 2\ \Omega + 2\ \Omega = 4\ \Omega$. Similarly, for choices 2, 3, and 4, $R_{eq} = 6\ \Omega$, $1\ \Omega$ and $2\ \Omega$ respectively. Choice 3 has the smallest equivalent resistance.

2. 3 Under Electricity - Series Circuits, find the equation $R_{eq} = R_1 + R_2 + R_3 + \ldots$. Substitution, gives $R_{eq} = X\ \Omega + X\ \Omega + X\ \Omega$. Solving, $R_{eq} = 3X\ \Omega$.

3. 4 Under Electricity - Series Circuits, find the equation $R_{eq} = R_1 + R_2 + R_3 + \ldots$. Since R_{eq} is the sum of the individual resistances, the total resistance must be greater than the largest resistance.

4. 4 The resistors are connected in series in the circuit. Under Electricity -Series Circuits, find the equation $R_{eq} = R_1 + R_2 + R_3 + \ldots$. Substitution and solving gives $R_{eq} = 8.0\ \Omega + 8.0\ \Omega = 16\ \Omega$. Now under Electricity in the reference table, find the equation $R = V/I$. Substitution gives $16\ \Omega = (12\ V)/(I)$. Solving, $I = 0.75\ A$.

5. 2 Under Electricity, find the equation $R = V/I$. Solving for the total resistance of the circuit gives $R = (12\ V)/(1.0\ A) = 12\ \Omega$. Since the two resistors are connected in series, use the equation $R_{eq} = R_1 + R_2 + R_3 + \ldots$, found under Series Circuits in the reference table. Substitution gives $12\ \Omega = 3.0\ \Omega + R_2$. Solving gives $R_2 = 9.0\ \Omega$.

6. 1 Using $R = V/I$ to solve for the current through the 20. Ω resistor. Substitution gives 20. $\Omega = (80.\ V)/(I)$. Solving, $I = 4.0\ A$. Since the resistors are connected in series, the current through the 40. Ω resistor is 4.0 A. Use the same equation to find V_1. Substitution gives 40. $\Omega = V/(4.0\ A)$. Solving, $V = 160\ V$.

7. 4 The series equation $R_{eq} = R_1 + R_2 + R_3 + \ldots$. shows that as the number of resistors in series increases, the total resistance increases. Under Electricity, now find the equation $R = V/I$. Solving, $I = V/R$. This shows that if the potential difference remains constant, as the resistance increases, the current decreases.

8. Answer: 6.2 J

   Explanation: Under Electricity, find the equation $W = V^2 t/R$, where R is the equivalent resistance of the circuit. The three resistors are connected in series. Using the resistance relationship under Electricity - Series Circuits, $R_{eq} = 8.0\ \Omega + 8.0\ \Omega + 10.\ \Omega = 26\ \Omega$. Substitution into the first equation gives $W = (4.0\ V)^2 (10.\ s)/(26\ \Omega)$. Solving, $W = 6.2\ J$.

Parallel Circuits: These circuits provide two or more separate paths for current.

The total circuit current (*I*) is equal to the sum of the individual component or branch currents.

$$I = I_1 + I_2 + I_3 + \dots$$ *where: I = current*

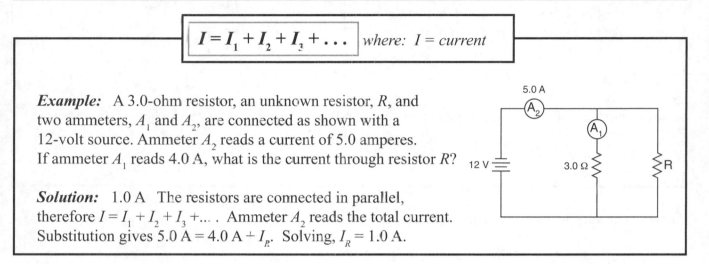

Example: A 3.0-ohm resistor, an unknown resistor, *R*, and two ammeters, A_1 and A_2, are connected as shown with a 12-volt source. Ammeter A_2 reads a current of 5.0 amperes. If ammeter A_1 reads 4.0 A, what is the current through resistor *R*?

Solution: 1.0 A The resistors are connected in parallel, therefore $I = I_1 + I_2 + I_3 + \dots$. Ammeter A_2 reads the total current. Substitution gives 5.0 A = 4.0 A + I_R. Solving, I_R = 1.0 A.

Since all the components are connected between two points whose potential difference is *V*, the potential difference across each component is the same as *V*.

$$V = V_1 = V_2 = V_3 = \dots$$ *where: V = potential difference*

Example: In the accompanying circuit diagram, what are the correct readings of voltmeters V_1 and V_2?

(1) V_1 reads 2.0 V and V_2 reads 4.0 V
(2) V_1 reads 4.0 V and V_2 reads 2.0 V
(3) V_1 reads 3.0 V and V_2 reads 3.0 V
(4) V_1 reads 6.0 V and V_2 reads 6.0 V

Solution: 4 The resistors are connected in parallel with each other. Therefore, $V_1 = V_2 = V = 6.0$ V

The reciprocal of the total or equivalent resistance of resistors in parallel is equal to the sum of the reciprocals of the individual resistances.

$$\frac{1}{R_{eq}} = \frac{1}{R_1} + \frac{1}{R_2} + \frac{1}{R_3} \ldots$$ *where:* R_{eq} = *equivalent resistance*
   R = *resistance*

Example: A 20.-ohm resistor and a 30.-ohm resistor are connected in parallel to a 12-volt battery as shown. What is the equivalent resistance of the circuit?

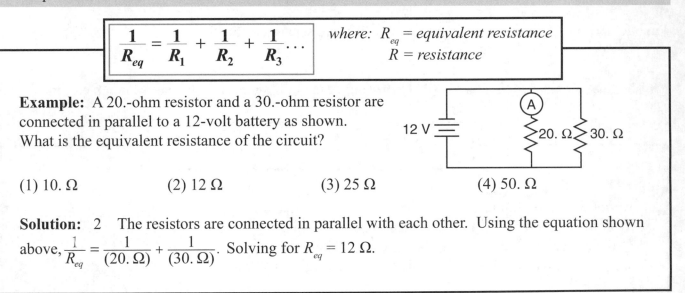

(1) 10. Ω (2) 12 Ω (3) 25 Ω (4) 50. Ω

Solution: 2   The resistors are connected in parallel with each other. Using the equation shown above, $\frac{1}{R_{eq}} = \frac{1}{(20.\,\Omega)} + \frac{1}{(30.\,\Omega)}$. Solving for $R_{eq} = 12\,\Omega$.

Parallel Circuits – Additional Information:

- $I = I_1 + I_2 + I_3 + \ldots$ This is the Law of Conservation of Charge applied to a parallel circuit.

- $V = V_1 = V_2 = V_3 = \ldots$ This is the Law of Conservation of energy applied to a parallel circuit.

- Because of the reciprocal relationship for equivalent resistance, the equivalent resistance of resistors in a parallel circuit must be less than the smallest of the individual resistors.

1. As the number of resistors in a parallel circuit is increased, what happens to the equivalent resistance of the circuit and total current in the circuit?

   (1) Both equivalent resistance and total current decrease.
   (2) Both equivalent resistance and total current increase.
   (3) Equivalent resistance decreases and total current increases.
   (4) Equivalent resistance increases and total current decreases.   1 _____

2. An electric circuit contains an operating heating element and a lit lamp. Which statement best explains why the lamp remains lit when the heating element is removed from the circuit?

   (1) The lamp has less resistance than the heating element.
   (2) The lamp has more resistance than the heating element.
   (3) The lamp and heating element were connected in series.
   (4) The lamp and heating element were connected in parallel.   2 _____

3. Two identical resistors connected in parallel have an equivalent resistance of 40. ohms. What is the resistance of each resistor?

   (1) 20. Ω   (3) 80. Ω
   (2) 40. Ω   (4) 160 Ω   3 _____

4. The diagram below represents part of an electric circuit containing three resistors.

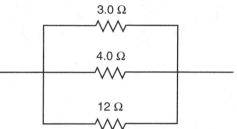

   What is the equivalent resistance of this part of the circuit?

   (1) 0.67 Ω   (3) 6.3 Ω
   (2) 1.5 Ω   (4) 19 Ω   4 _____

Base your answers to questions 5 and 6 on the diagram below, which represents an electric circuit consisting of four resistors and a 12-volt battery.

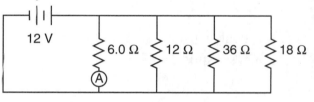

5. What is the current measured by ammeter A?

   (1) 0.50 A   (3) 72 A
   (2) 2.0 A   (4) 4.0 A   5 _____

6. What is the equivalent resistance of this circuit?

   (1) 72 Ω   (3) 3.0 Ω
   (2) 18 Ω   (4) 0.33 Ω   6 _____

Base your answers to questions 7a, b and c
on the information and accompanying diagram.

A 3.0-ohm resistor, an unknown resistor, R, and two
ammeters, A_1 and A_2, are connected as shown with a
12-volt source. Ammeter A_2 reads a current of 5.0 amperes.

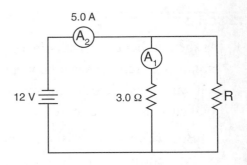

7. a) Determine the equivalent resistance of the circuit.

    b) Calculate the current measured by ammeter A_1. [Show all work, including the equation and
       substitution with units.]

    c) Calculate the resistance of the unknown resistor, R. [Show all work, including the equation
       and substitution with units.]

Base your answers to questions 8a and b
on the accompanying circuit diagram, which
shows two resistors connected to a 24-volt
source of potential difference.

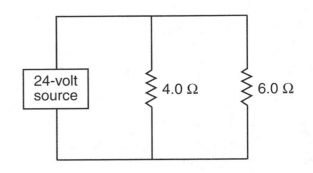

8. a) On the accompanying diagram, use the
       appropriate circuit symbol to indicate a
       correct placement of a voltmeter to determine
       the potential difference across the circuit.

    b) What is the total resistance of the circuit? _____

Copyright © 2011
Topical Review Book Company

Base your answers to questions 9a, b and c on the information and diagram below.

A 15-ohm resistor, R_1, and a 30.-ohm resistor, R_2, are to be connected in parallel between points A and B in a circuit containing a 90.-volt battery.

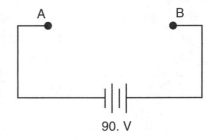

9. a) Complete the accompanying diagram to show the two resistors connected in parallel between points A and B.

    b) Determine the potential difference across resistor R_1.

    c) Calculate the current in resistor R_1. [Show all work, including the equation and substitution with units.]

Base your answers to question 10a through d on the information and data table below.

Three lamps were connected in a circuit with a battery of constant potential. The current, potential difference, and resistance for each lamp are listed in the data table below. [There is negligible resistance in the wires and the battery.]

	Current (A)	Potential Difference (V)	Resistance (Ω)
lamp 1	0.45	40.1	89
lamp 2	0.11	40.1	365
lamp 3	0.28	40.1	143

10. a) What is the potential difference supplied by the battery? _____ V

    b) Calculate the equivalent resistance of the circuit. [Show all work, including the equation and substitution with units.]

    c) If lamp 3 is removed from the circuit, what would be the value of the potential difference across lamp 1 after lamp 3 is removed? _____ V

    d) If lamp 3 is removed from the circuit, what would be the value of the current in lamp 2 after lamp 3 is removed? _____ A

Note: Question 11 has only three choices.

11. In the diagram below, lamps L_1 and L_2 are connected to a constant voltage power supply.

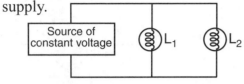

If lamp L_1 burns out, the brightness of L_2 will

(1) decrease

(2) increase

(3) remain the same   11 _____

12. The diagram below, shows two resistors and three ammeters connected to a voltage source.

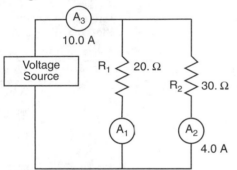

What is the current reading of ammeter A_1?

(1) 10.0 A   (3) 3.0 A

(2) 6.0 A   (4) 4.0 A   12 _____

13. What is the total resistance of the circuit segment shown in the diagram below?

(1) 1.0 Ω

(2) 9.0 Ω

(3) 3.0 Ω

(4) 27 Ω

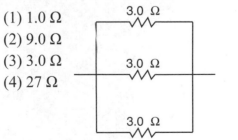

13 _____

Base your answers to questions 14 and 15 on the circuit diagram below.

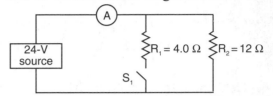

14. If switch S_1 is open, the reading of ammeter A is

(1) 0.50 A   (3) 1.5 A

(2) 2.0 A   (4) 6.0 A   14 _____

15. If switch S_1 is closed, the equivalent resistance of the circuit is

(1) 8.0 Ω   (3) 3.0 Ω

(2) 2.0 Ω   (4) 16 Ω   15 _____

16. In the circuit diagram shown below, ammeter A_1 reads 10. amperes.

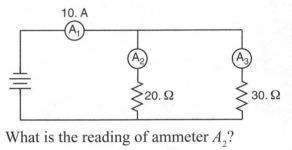

What is the reading of ammeter A_2?

(1) 6.0 A   (3) 20. A

(2) 10. A   (4) 4.0 A   16 _____

17. Three resistors, 4 ohms, 6 ohms, and 8 ohms, are connected in parallel in an electric circuit. The equivalent resistance of the circuit is

(1) less than 4 Ω

(2) between 4 Ω and 8 Ω

(3) between 10. Ω and 18 Ω

(4) 18 Ω   17 _____

18. An electric circuit contains two 3.0-ohm resistors connected in parallel with a battery. The circuit also contains a voltmeter that reads the potential difference across one of the resistors.

Calculate the total resistance of the circuit. [Show all work, including the equation and substitution with units.]

Base your answers to questions 19a, b, and c on the information below.

An 18-ohm resistor and a 36-ohm resistor are connected in parallel with a 24-volt battery. A single ammeter is placed in the circuit to read its total current.

19. a) In the space below, draw a diagram of this circuit using symbols from the Reference Tables for Physical Setting/Physics. [Assume the availability of any number of wires of negligible resistance.]

b) Calculate the equivalent resistance of the circuit. [Show all work, including the equation and substitution with units.]

c) Calculate the total power dissipated in the circuit. [Show all work, including the equation and substitution with units.]

20. A 5.0-ohm resistor, a 10.0-ohm resistor, and a 15.0-ohm resistor are connected in parallel with a battery. The current through the 5.0-ohm resistor is 2.4 amperes. A 20.0-ohm resistor is added to the circuit in parallel with the other resistors. Describe the effect the addition of this resistor has on the amount of electrical energy expended in the 5.0-ohm resistor in 2.0 minutes.

_____

_____

1. **3** Under Electricity, find the equation $1/R_{eq} = 1/R_1 + 1/R_2 + 1/R_3 + ...$. Since this is a reciprocal equation, as the number of resistors in parallel increases, the equivalent resistance decreases. From Ohm's Law, current varies inversely with resistance under constant voltage. Therefore, as the resistance decreases, the current increases.

2. **4** Since the lamp remains lit when the heating element is removed from the circuit, each must provide its own path for current. Therefore, they must be connected in parallel with each other.

3. **3** Under Electricity - Parallel Circuits, find the equation $1/R_{eq} = 1/R_1 + 1/R_2 + 1/R_3 + ...$. Let $R_1 = R_2 = R$. Then $1/40. \ \Omega = 1/R + 1/R$. Solving for R gives $80. \ \Omega$.

4. **2** The resistors in the circuit are connected in parallel. Under Electricity - Parallel Circuits, find the equation $1/R_{eq} = 1/R_1 + 1/R_2 + 1/R_3 + ...$. Substitution into the equation gives $1/R_{eq} = 1/(3.0 \ \Omega) + 1/(4.0 \ \Omega) + 1/(12 \ \Omega)$. Solving gives $R_{eq} = 1.5 \ \Omega$.

5. **2** The resistors are connected in parallel to the 12 V source. Therefore, the potential difference across each must be 12 V (see the equation $V = V_1 = V_2 = V_3 = ...$ under Electricity - Parallel Circuits). Under Electricity, locate the equation $R = V/I$. Substitution gives $6.0 \ \Omega = (12 \ V)/(I)$. Solving, $I = 2.0 \ A$.

6. **3** Under Electricity - Parallel Circuits, find the equation $1/R_{eq} = 1/R_1 + 1/R_2 + 1/R_3 + ...$. Substitution gives $1/R_{eq} = 1/(6.0 \ \Omega) + 1/(12 \ \Omega) + 1/(36 \ \Omega) + 1/(18 \ \Omega)$. Solving, $R_{eq} = 3.0 \ \Omega$.

7. *a)* Answer: $2.4 \ \Omega$

    Explanation: Under Electricity in the reference table, find the equation $R = V/I$ (Ohm's Law). The equivalent resistance of the circuit can be calculated using the total circuit voltage (12 V) and total circuit current (5.0 A). Substitution gives $R = \dfrac{(12 \ V)}{(5.0 \ A)}$ and $R = 2.4 \ \Omega$.

    b) $R = \dfrac{V}{I}$

    $I = \dfrac{V}{R}$

    $I = \dfrac{12 \ V}{3.0 \ \Omega}$

    $I = 4.0 \ A$

    Explanation: Since the two resistors are connected in parallel to the voltage source, the voltage across each is 12 V (see the equation $V = V_1 = V_2 = V_3 = ...$ Under Electricity – Parallel Circuits).

    c) $R = \dfrac{V}{I}$ $\qquad \dfrac{1}{R_{eq}} = \dfrac{1}{R_1} + \dfrac{1}{R_2}$

    $R = \dfrac{12 \ V}{1.0 \ A}$ *or* $\dfrac{1}{2.4 \ \Omega} = \dfrac{1}{3.0 \ \Omega} + \dfrac{1}{R_2}$

    $R = 12 \ \Omega$ $\qquad \qquad R = 12 \ \Omega$

    Explanation: Under Electricity – Parallel Circuits, find the equation $I = I_1 + I_2 + I_3 + ...$ I represents the total circuit current, 5.0 A. From question 7b, we know the current through the 3.0 Ω resistor is 4.0 A. The current through R must then be 5.0 A = 4.0 A + I_R or 1.0 A. Knowing the voltage across R is 12 V (see question 7b), use Ohm's Law to calculate R

8. a)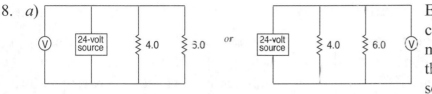

Explanation: A voltmeter must be connected in parallel with a device to measure the potential difference. Since the resistors are connected to the 24 V source in parallel, the measured potential difference across any resistor will equal the potential difference of the circuit ($V = V_1 = V_2$).

b) 2.4 Ω   Explanation: The resistors are connected in parallel with each other. Under Electricity - Parallel Circuits, find the equation $1/R_{eq} = 1/R_1 + 1/R_2 + 1/R_3 + ...$. Substitution into the equation gives $1/R_{eq} = 1/4\ \Omega + 1/6\ \Omega = 2.4\ \Omega$ for the total resistance.

9. a)

Explanation: When connected in parallel, the resistors provide separate paths for current.

b) Answer: 90.0 V

Explanation: Under Electricity - Parallel Circuits, find the equation $V = V_1 = V_2 = V_3 = ...$. This equation indicates that the potential difference across each resistor is equal to the potential difference of the source (90. V).

c) $R = \dfrac{V}{I}$

$I = \dfrac{V}{R}$

$I = \dfrac{90.\ V}{15\ \Omega}$

$I = 6.0\ A$

Explanation: Under Electricity, find the equation $R = V/I$. Using the potential difference across R from answer 9b, $15\ \Omega = (90.\ V)/I$. Solving, $I = 6.0\ A$.

10. a) 40.1 V   Explanation: Under Electricity - Parallel Circuits, the equation for the potential difference shows that the potential difference across each device is equal to the potential difference of the source.

b) $1/R_{eq} = 1/R_1 + 1/R_2 + 1/R_3 = 1/89\ \Omega + 1/265\ \Omega + 1/143\ \Omega$ $R_{eq} = 48\ \Omega$ (or 47.7 Ω)

or $I = I_1 + I_2 + I_3 = 0.45\ A + 0.11\ A + 0.28\ A = 0.84\ A$ $R = V/I = 40.1\ V/0.84\ A = 48\ \Omega$

Explanation: Under Electricity - Parallel Circuits, find the equation $1/R_{eq} = 1/R_1 + 1/R_2 + 1/R_3$. Substitute the values for the resistances from the table into the equation and solve for R_{eq}.
or   Calculate the total circuit current using the current relationship under Parallel Circuits ($I = I_1 + I_2 + I_3$) and use $R = V/I$ to calculate R.

c) The potential difference is 40.1 V

Explanation: In a parallel circuit, removal of one device does not affect the other devices since each is in its own conducting path. Removal of lamp 3 therefore will not affect lamp 1. The potential difference across lamp 1 remains 40.1 V.

d) 0.11 A   Explanation: Using the same reasoning as in 10c, removal of lamp 3 will not affect lamp 2. The current through lamp 2 remains 0.11 A.

Set 1 — Parallel and Series Circuits

1. Two identical resistors connected in series have an equivalent resistance of 4 ohms. The same two resistors, when connected in parallel, have an equivalent resistance of

   (1) 1 Ω   (3) 8 Ω
   (2) 2 Ω   (4) 4 Ω   1 _____

2. Which circuit diagram below correctly shows the connection of ammeter A and voltmeter V to measure the current through and potential difference across resistor R?

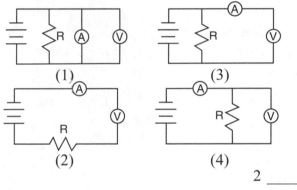

   2 _____

3. In which circuit would ammeter A show the greatest current?

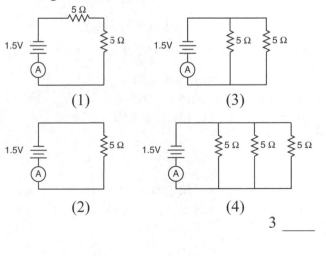

   3 _____

4. A physics student is given three 12-ohm resistors with instructions to create the circuit that would have the lowest possible resistance. The correct circuit would be a

   (1) series circuit with an equivalent resistance of 36 Ω
   (2) series circuit with an equivalent resistance of 4.0 Ω
   (3) parallel circuit with an equivalent resistance of 36 Ω
   (4) parallel circuit with an equivalent resistance of 4.0 Ω   4 _____

5. In which circuit represented below are meters properly connected to measure the current through resistor R_1 and the potential difference across resistor R_2?

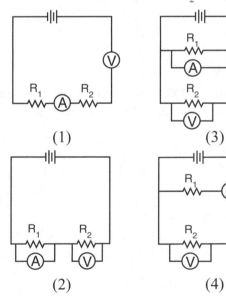

   5 _____

6. In which circuit would current flow through resistor R_1, but not through resistor R_2 while switch S is open?

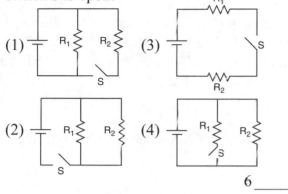

(1)   (3)

(2)   (4)

6 _____

7. When a 15-ohm resistor is connected in parallel with a 30-.ohm resistor, the equivalent resistance is 10.-ohm. If these two resistors are connected in series the equivalent resistance is

(1) 10 Ω   (3) 30 Ω
(2) 15 Ω   (4) 45 Ω   7 _____

8. In which pair of circuits shown below could the readings of voltmeters V_1 and V_2 and ammeter A be correct?

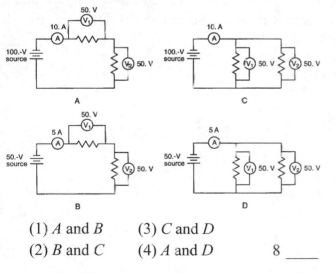

(1) A and B   (3) C and D
(2) B and C   (4) A and D   8 _____

9. A 6.0-ohm lamp requires 0.25 ampere of current to operate. In which circuit below would the lamp operate correctly when switch S is closed?

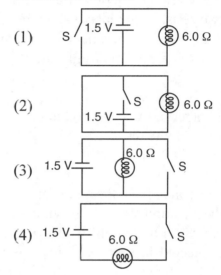

(1)
(2)
(3)
(4)

9 _____

10. Identical resistors (R) are connected across the same 12-volt battery. Which circuit uses the greatest power?

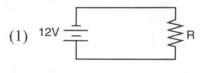
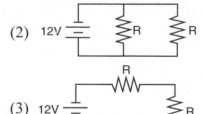

(1) 12V   R
(2) 12V   R   R
(3) 12V   R ... R
(4) 12V   R R R R

10 _____

Answers

Set 1

1. 1 Under Electricity – Series Circuits, in the reference table, find the equation $R_{eq} = R_1 + R_2 + R_3 + \ldots$. The two resistors are identical, therefore $R_1 = R_2$. In series, $4\,\Omega = R_1 + R_2$ and $R_1 = R_2 = 2\,\Omega$. Under Electricity – Parallel Circuits, find the equation $1/R_{eq} = 1/R_1 + 1/R_2 + 1/R_3 + \ldots$. Using a value of $2\,\Omega$ for R_1 and R_2 gives $1/R_{eq} = 1/2\,\Omega + 1/2\,\Omega$ and $R_{eq} = 1\,\Omega$.

2. 4 An ammeter must be connected in series with the resistor and a voltmeter must be connected in parallel with the resistor.

3. 4 In the circuits given, the ammeter is connected to show the total circuit current. Under Electricity, find the equation $R = V/I$. Solving for I, $I = V/R$. At the same potential difference, the current will be greatest in the circuit with the smallest equivalent resistance. In choice 1, the two resistors are connected in series. Using the resistance relationship under Electricity - Series circuits, $R_{eq} = 5\,\Omega + 5\,\Omega = 10\,\Omega$. In choice 2, the total resistance is $5\,\Omega$. In choices 3 and 4, the resistors are connected in parallel. Using the resistance relationship under Electricity - Parallel Circuits, $1/R_{eq} = 1/5\,\Omega + 1/5\,\Omega$ ($R_{eq} = 2.5\,\Omega$) and $1/R_{eq} = 1/5\,\Omega + 1/5\,\Omega + 1/5\,\Omega$ ($R_{eq} = 0.6\,\Omega$) for choices 3 and 4 respectively. Choice 4, having the smallest equivalent resistance would therefore have the greatest current.

4. 4 Under Electricity - Series Circuits, find the equation $R_{eq} = R_1 + R_2 + R_3 + \ldots$. When connected in series, $R_{eq} = 12\,\Omega + 12\,\Omega + 12\,\Omega = 36\,\Omega$. Under Electricity - Parallel Circuits, find the equation $1/R_{eq} = 1/R_1 + 1/R_2 + 1/R_3 + \ldots$. When connected in parallel, $1/R_{eq} = 1/12\,\Omega + 1/12\,\Omega + 1/12\,\Omega$ and $R_{eq} = 4\,\Omega$.

5. 4 An ammeter must be connected in series with a resistor to measure the current through it and a voltmeter must be connected in parallel with a resistor to measure the potential difference across it.

Charts – List of Physical Constants

List of Physical Constants		
Name	**Symbol**	**Value**
Universal gravitational constant	G	6.67×10^{-11} N•m²/kg²
Acceleration due to gravity	g	9.81 m/s²
Speed of light in a vacuum	c	3.00×10^{8} m/s
Speed of sound in air at STP		3.31×10^{2} m/s
Mass of Earth		5.98×10^{24} kg
Mass of the Moon		7.35×10^{22} kg
Mean radius of Earth		6.37×10^{6} m
Mean radius of the Moon		1.74×10^{6} m
Mean distance—Earth to the Moon		3.84×10^{8} m
Mean distance—Earth to the Sun		1.50×10^{11} m
Electrostatic constant	k	8.99×10^{9} N•m²/C²
1 elementary charge	e	1.60×10^{-19} C
1 coulomb (C)		6.25×10^{18} elementary charges
1 electronvolt (eV)		1.60×10^{-19} J
Planck's constant	h	6.63×10^{-34} J•s
1 universal mass unit (u)		9.31×10^{2} MeV
Rest mass of the electron	m_e	9.11×10^{-31} kg
Rest mass of the proton	m_p	1.67×10^{-27} kg
Rest mass of the neutron	m_n	1.67×10^{-27} kg

Overview:

In solving many problems in Physics, one needs the value and units of a number that is very specific to that problem. These values and units are given on this table.

The Table:

This table gives the Name, Symbol and Value of many of the fundamental constants in Physics. Those without a symbol listed are constants used in many problems or as conversion factors in changing from one unit to another for a given quantity.

The coulomb (C), electronvolt (eV) and universal mass unit (u) and the respective values given are used as conversion factors.

The information about the Earth, Moon and Sun are used primarily in problems involving the gravitational forces between these objects.

The speed of sound in air at STP may simply be represented by the symbol v in equations.

Additional Information:

- The speed of light in a vacuum (c) is the speed of all electromagnetic waves or radiation in a vacuum (see The Electromagnetic Spectrum).

- Planck's Constant (h - the quantum constant) is extremely small. For this reason, quantum effects are not noticeable in everyday life.

- The universal mass unit (u) gives the relationship between mass and energy.

- The value of the acceleration due to gravity (g) decreases as one moves above the Earth's surface but may be considered constant near the Earth's surface.

- The elementary charge (e) is the minimum charge found on a particle in nature and may be positive or negative. It is the charge of a proton or an electron.

- Of the two forces most noticeable in everyday life, gravitational and electrical, gravity is the weakest. Notice the very small value of the universal gravitational constant compared to the electrostatic constant.

Questions dealing with **List of Physical Constants** will be found throughout this workbook.

Charts

Charts – Prefixes for Powers of 10

Prefixes for Powers of 10		
Prefix	**Symbol**	**Notation**
tera	T	10^{12}
giga	G	10^{9}
mega	M	10^{6}
kilo	k	10^{3}
deci	d	10^{-1}
centi	c	10^{-2}
milli	m	10^{-3}
micro	μ	10^{-6}
nano	n	10^{-9}
pico	p	10^{-12}

Overview:

In physics, one uses many numbers that are very small or very large. The system of measurement used in the sciences is the Metric System, also referred to as the SI system. In addition to the defined standards for the various quantities, amounts that are multiples or submultiples of 10 are used. Greek numerical prefixes are frequently used to express multiples or submultiples of these standards when they are very small or very large.

The Table:

This table lists the Prefix, Symbol and Notation (in exponential form) of the various multiples and submultiples of 10 frequently used. By knowing the exponent for a certain prefix, as shown in the Notation column, one can determine how many places the decimal must be moved when performing conversions. For example, as shown in the Notation column, changing from kilo- to milli- (or vice versa) requires a six place decimal movement. For most everyday activities, the most common prefixes used are kilo-, centi- and milli-.

Additional Information:

- The defined base quantities and the units in the Metric System are those of length (the meter - m), mass (the kilogram - kg) and time (the second - s).

- The capacity of memory chips for electronic devices (computers, Ipods, digital cameras) are expressed in megabytes (10^6 or one million bytes) or gigabytes (10^9 or one billion bytes).

- The speed at which a computer completes an operation is expressed in nanoseconds (ns), which are billionths (10^{-9}) seconds.

Set 1 — Prefixes For Powers of Ten

1. 1.78 km = ______________________ m

2 0.00076 m = ______________________ cm

3. 0.0000000065 m = ______________________ nm

4. 2,500 J = ______________________ kJ

5. 8,400,000 μA = ______________________ A

6. 2,000,000,000,000 V = ______________________ TV

7. 7,650,000 g = ______________________ Mg

8. 45,600 kJ = ______________________ J

9. 0.000000082 s = ______________________ ns

10. 1.2 A = ______________________ mA

11. 1.2 G Ω = ______________________ Ω

12. 45,600,000,000 Hz = ______________________ GHz

13. 3 TV = ______________________ V

14. 5.6 MW = ______________________ W

15. 2.54 μg = ______________________ g

16. 5.6 nm = ______________________ m

17. 0.00000004 s = ______________________ μs

18. The energy produced by the complete conversion of 2.0×10^{-5} kilograms of mass into energy is 1.8×10^{12} J. This is equal to

    (1) 1.8 TJ   (3) 1.8 MJ

    (2) 1.8 GJ   (4) 1.8 kJ   18 _____

19. As shown in the List of Physical Constants, the mean radius of the Moon is

    (1) 1.74×10^2 Mm

    (2) 17.4 Mm

    (3) 1.74 Mm

    (4) 1.74×10^{-1} Mm   19 _____

20. As shown in the List of Physical Constants, the mean distance – Earth to the Sun is

    (1) 1.50×10^2 Tm

    (2) 1.50×10^{-1} Tm

    (3) 1.50×10^{-2} Tm

    (4) 1.50×10^{-3} Tm   20 _____

21. Which quantity is equivalent to 5,000 kilocalories?

    (1) 0.5 megacalories

    (2) 5 megacalories

    (3) 50 megacalories

    (4) 500 megacalories   21 _____

22. An Arizona solar power plant is able to produce 24 megawatts, which can serve 6,000 homes. This averages out to be how many watts per home?

    (1) 40 W   (3) 4,000 W

    (2) 400 W   (4) 40,000 W   22 _____

23. The polarity of AC current reverses every 20 milliseconds. This time period is equal to

    (1) 0.02 s   (3) 0.000002 s

    (2) 0.002 s   (4) 2000 s   23 _____

24. 2.6×10^{-6} joules is how many microjoules? _____ μJ

25. 5,600 milligrams is equal to how many grams? _____ g

26. 0.089 microseconds is equal to how many nanoseconds? _____ ns

27. 0.0056 gigavolts is equal to how many megavolts? _____ MV

28. 0.00082 nanoseconds equals how many picoseconds? _____ ps

29. 35,000 m can be expressed as _____ km

30. 13,800 joules per gram can be expressed as _____ kJ/g

31. 6,700 mA can be expressed as _____ A

Prefixes for Powers of 10
Answers
Set 1

1. 1.78 km = _____ 1.78×10^3 _____ m

2. 0.00076 m = _____ 7.6×10^{-2} _____ cm

3. 0.0000000065 m = _____ 6.5 _____ nm

4. $2,500$ J = _____ 2.5 _____ kJ

5. $8,400,000$ μA = _____ 8.4 _____ A

6. $2,000,000,000,000$ V = _____ 2 _____ TV

7. $7,650,000$ g = _____ 7.65 _____ Mg

8. $45,600$ kJ = _____ 4.56×10^7 _____ J

9. 0.000000082 s = _____ 82×10^{-8} _____ ns

10. 1.2 A = _____ 1.2×10^3 _____ mA

11. 1.2 G Ω = _____ 1.2×10^9 _____ Ω

12. $45,600,000,000$ Hz = _____ 45.6 _____ GHz

13. 3 TV = _____ 3×10^{12} _____ V

14. 5.6 MW = _____ 5.6×10^6 _____ W

15. 2.54 μg = _____ 2.54×10^{-6} _____ g

16. 5.6 nm = _____ 5.6×10^{-9} _____ m

17. 0.00000004 s = _____ 4×10^{-2} _____ μs

Approximate Coefficients of Friction

Overview:

Friction is a force that opposes motion. This force is caused by irregularities in the surfaces of objects, electrical forces between objects and adhesive and cohesive forces between surfaces. The coefficient of friction (μ) is defined as the ratio of the force needed to overcome friction (the frictional force - F_f) to the force pressing the surfaces together (the normal force or the force perpendicular to the surfaces - F_N) or $\mu = F_f/F_N$. The force of friction is then $F_f = \mu F_N$. This equation is given under Mechanics.

Approximate Coefficients of Friction

	Kinetic	Static
Rubber on concrete (dry)	0.68	0.90
Rubber on concrete (wet)	0.58	
Rubber on asphalt (dry)	0.67	0.85
Rubber on asphalt (wet)	0.53	
Rubber on ice	0.15	
Waxed ski on snow	0.05	0.14
Wood on wood	0.30	0.42
Steel on steel	0.57	0.74
Copper on steel	0.36	0.53
Teflon on Teflon	0.04	

The Table:

This table gives the identity of the surfaces and the values of two types of coefficients of friction - Kinetic and Static. Kinetic friction, also known as sliding friction, is the force of friction encountered when one surface is sliding over another. Static friction, also known as starting friction, is the force encountered when starting the movement of one surface over another.

Since the coefficient of friction is a ratio of two forces, the coefficient has no units.

Be sure to read the question carefully and then choose the correct value (kinetic or static) for the coefficient of friction.

Additional Information:

- The force of friction depends upon the substances and their surfaces. Therefore, only the names of the substances or materials have to be known to choose the correct coefficient of friction.

- Within the range of medium speeds, the force of kinetic or sliding friction is nearly independent of the speed of sliding.

- The force of kinetic friction is nearly independent of the area of contact.

- Notice that the coefficient of kinetic friction is less than the coefficient of static friction. Thus, once one surface is moving over another surface, it requires less force to maintain the motion than it did to start the motion.

- The force of friction acts parallel to the surfaces which are moving over one another.

- Water acts as a lubricant when on the surface of a substance, reducing the force of friction between the surfaces. Hence, the coefficient of kinetic friction for rubber on concrete (wet) and rubber on asphalt (wet) is smaller than that for rubber on concrete (dry) and rubber on asphalt (dry).

Questions dealing with **Approximate Coefficients of Friction** will be found in the **Friction** section.

The Electromagnetic Spectrum

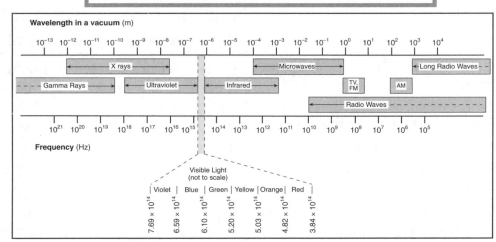

Overview:

Electromagnetic waves or radiation are transverse waves generated by accelerated charged particles. These waves consist of two transverse waves, one electric and the other magnetic, vibrating at a right angle to each other and to the direction of propagation.

The Electromagnetic Spectrum is a continuous range of electromagnetic waves or radiation ranging in frequency from about 10 to 10^{25} Hz. Wavelengths range from about 3×10^7 to 3×10^{-17} m.

The Table:

The table lists the Names, Wavelength range and Frequency range of the families of the electromagnetic spectrum. The families of the electromagnetic spectrum differ only in the frequency of the waves and their source. There is no sharp boundary between the families.

Additional Information:

- In an electromagnetic wave, the changing electric wave induces a changing magnetic wave and vice versa. Therefore, electromagnetic waves do not require a medium for transmission.

- In a vacuum (or air), all electromagnetic waves travel with a speed of 3.00×10^8 m/s. This is known as the speed of light (c). See the List of Physical Constants.

- The human eye is sensitive to only a very small range of electromagnetic waves known a visible light. This range is referred to as the visible spectrum, consisting the six spectral colors that are listed on the chart.

- As the frequency of the electromagnetic wave increases, the wavelength decreases.

- As the frequency of the electromagnetic waves increases, the wave nature of the radiation decreases and the particle nature increases.

- The infrared family is also known as heat waves. The ultraviolet family is referred to as black light.

- X-rays are used as diagnostic tools in medicine.

- Gamma rays are high energy rays emitted by radioactive elements.

- Microwaves cook the food in microwave ovens.

- Radio waves include TV, AM and FM signals.

1. Compared to the speed of a sound wave in air, the speed of a radio wave in air is

   (1) less
   (2) greater
   (3) the same   1 _____

2. Electromagnetic radiation having a wavelength of 1.3×10^{-7} meter would be classified as

   (1) infrared   (3) blue
   (2) orange   (4) ultraviolet   2 _____

3. Which pair of terms best describes light waves traveling from the Sun to Earth?

   (1) electromagnetic and transverse
   (2) electromagnetic and longitudinal
   (3) mechanical and transverse
   (4) mechanical and longitudinal   3 _____

4. Which wave characteristic is the same for all types of electromagnetic radiation traveling in a vacuum?

   (1) speed   (3) period
   (2) wavelength   (4) frequency   4 _____

5. A photon of which electromagnetic radiation has the most energy?

   (1) ultraviolet   (3) infrared
   (2) x ray   (4) microwave   5 _____

Base your answers to question 6 on the data table below. The data table lists the energy and corresponding frequency of five photons.

Photon	Energy (J)	Frequency (Hz)
A	6.63×10^{-15}	1.00×10^{19}
B	1.99×10^{-17}	3.00×10^{16}
C	3.49×10^{-19}	5.26×10^{14}
D	1.33×10^{-20}	2.00×10^{13}
E	6.63×10^{-26}	1.00×10^{8}

6. In which part of the electromagnetic spectrum would photon D be found?

   (1) infrared   (3) ultraviolet
   (2) visible   (4) x ray   6 _____

7. An electromagnetic AM-band radio wave could have a wavelength of

   (1) 0.005 m   (3) 500 m
   (2) 5 m   (4) 5 000 000 m   7 _____

8. A microwave and an x ray are traveling in a vacuum. Compared to the wavelength and period of the microwave, the x ray has a wavelength that is

   (1) longer and a period that is shorter
   (2) longer and a period that is longer
   (3) shorter and a period that is longer
   (4) shorter and a period that is shorter

                                          8 _____

9. A monochromatic beam of light has a frequency of 6.5×10^{14} hertz. What color is the light?

   (1) yellow   (3) violet
   (2) orange   (4) blue   9 _____

Base your answers to questions 10 *a* and *b* on the information below and the accompanying graph.

Sunlight is composed of various intensities of all frequencies of visible light. The graph represents the relationship between light intensity and frequency.

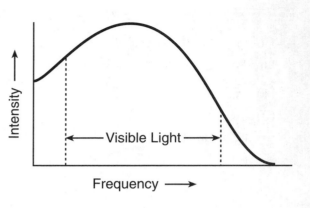

10. *a*) Based on the graph, which color of visible light has the lowest intensity?

_____

b) It has been suggested that fire trucks be painted yellow green instead of red. Using information from the graph, explain the advantage of using yellow-green paint.

_____

_____

_____

_____

11. The accompanying diagram represents a transverse wave traveling to the right through a medium. Point *A* represents a particle of the medium. At position *A* draw an arrow showing the direction that particle *A* will move in the next instant of time?

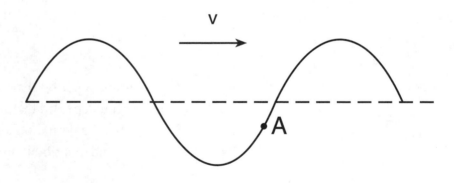

12. Which type of wave requires a material medium through which to travel?

    (1) radio wave
    (2) microwave
    (3) light wave
    (4) mechanical wave   12 _____

13. Which characteristic is the same for every color of light in a vacuum?

    (1) energy   (3) speed
    (2) frequency   (4) period   13 _____

14. Which wavelength is in the infrared range of the electromagnetic spectrum?

    (1) 100 nm   (3) 100 m
    (2) 100 mm   (4) 100 μm   14 _____

15. What is the speed of a radio wave in a vacuum?

    (1) 0 m/s
    (2) 3.31×10^2 m/s
    (3) 1.13×10^3 m/s
    (4) 3.00×10^8 m/s   15 _____

16. Compared to the wavelength of red light, the wavelength of yellow light is

    (1) shorter
    (2) longer
    (3) the same   16 _____

17. As a transverse wave travels through a medium, the individual particles of the medium move

    (1) perpendicular to the direction of wave travel
    (2) parallel to the direction of wave travel
    (3) in circles
    (4) in ellipses   17 _____

Note: Question 18 has only three choices.

18. Compared to the speed of microwaves in a vacuum, the speed of x-rays in a vacuum is

    (1) less   (2) greater   (3) the same   18 _____

19. Which color of light has a wavelength of 5.0×10^{-7} meter in air?

    (1) blue   (3) orange
    (2) green   (4) violet   19 _____

20. Give the properties of a transverse wave.

    _____

    _____

1. 2   Both light and radio waves are part of the electromagnetic spectrum and have the same speed in air (which is the same speed in a vacuum). Both speeds are found in the List of Physical Constants. Here it shows that electromagnetic waves are much faster than the speed of sound.

2. 4   On The Electromagnetic Spectrum chart a wavelength of 1.3×10^{-7} m places this radiation in the ultraviolet family.

3. 1   Light is a part of the electromagnetic spectrum (see The Electromagnetic Spectrum). All electromagnetic waves are transverse waves.

4. 1   The speed of all electromagnetic waves traveling in a vacuum is the same (the speed of light, c). As seen on The Electromagnetic Spectrum chart, the families of electromagnetic waves differ in frequency and wavelength.

5. 2   The higher the frequency, the greater the energy of the photon. Referring to The Electromagnetic Spectrum chart, x-ray radiation has the highest frequency of the choices given.

6. 1   Find the chart of The Electromagnetic Spectrum. The frequency of photon D falls in the infrared (IR) portion of the spectrum.

7. 3   Open to The Electromagnetic Spectrum chart. The AM-band radio waves have a wavelength range of 10^2 to 10^3 m.

8. 4   Find The Electromagnetic Spectrum chart. The wavelength of an x-ray is shown to be shorter than the wavelength of a microwave. The frequency of the x-ray is higher than the frequency of the microwave. Under Waves, find the equation $T = 1/f$. This indicates that the higher the frequency, the shorter the period.

9. 4   In The Electromagnetic Spectrum, the light spectrum is enlarged. Here it shows that a light beam having a frequency of 6.5×10^{-14} Hz is located in the blue area.

10. *a*) violet   *or*   the one with the greatest frequency

Explanation: According to the graph, the color of visible light with the lowest intensity is the color with the highest frequency. Referring to The Electromagnetic Spectrum chart, the color of visible light with the highest frequency is violet.

b) Yellow green has a higher intensity.   *or*   Yellow green is brighter than red.

Explanation: According to the graph, the color of visible light with the greatest intensity is in the middle of the frequency range of visible light. From The Electromagnetic Spectrum chart, this is in the yellow-green region.

11.

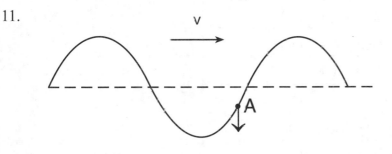

Point *A* moves up and down as the wave passes. It does not move along with the wave. As the wave moves to the right, the bottom of a trough is approaching point *A*. Therefore, point *A* is moving down.

Absolute Indices of Refraction $(f = 5.09 \times 10^{14}$ Hz)	
Air	1.00
Corn oil	1.47
Diamond	2.42
Ethyl alcohol	1.36
Glass, crown	1.52
Glass, flint	1.66
Glycerol	1.47
Lucite	1.50
Quartz, fused	1.46
Sodium chloride	1.54
Water	1.33
Zircon	1.92

Overview:

The speed of light in a material medium depends upon the properties of the medium. The absolute index of refraction (n) is defined as the ratio of the speed of light in a vacuum (c) to the speed of light in a material medium (v). Thus, $n = c/v$. The speed of light in a vacuum is found on the List of Physical Constants and the equation is found in the reference table under Waves.

The Table:

The table gives the value of the absolute index of refraction for various transparent material mediums. Since the index of refraction depends upon the frequency or color of the light, the values are given for a frequency of 5.09×10^{14} m, which is yellow light (see The Electromagnetic Spectrum). The chart shows that the indices of refraction have a range from 1.00 for air to 2.42 for diamond. As a ray of light enters a medium with a higher index of refraction, its speed decreases and it is refracted toward the normal. As a ray of light enters a medium with a smaller index of refraction, its speed increases and it is refracted away from the normal. Since the index of refraction is the ratio of two speeds, the index of refraction does not have a unit.

Note: For more detail, see the equation section under **Waves – Refraction**.

Additional Information:

- Expressed to three significant figures, the speed of light in air is the same as the speed of light in a vacuum. Therefore, the index of refraction of air is 1.00.

- From the equation for the absolute index of refraction, $v = c/n$. Since c is a constant, the speed of light in a medium varies inversely with the index of refraction. Therefore, as the value of the index of refraction increases, the speed of light in that medium decreases.

- Since the speed of light in a material medium is less than that in a vacuum, the index of refraction of any material medium must be greater than 1.

- The absolute index of refraction is a characteristic of that medium and may be used to identify that material.

- Diamond has a very large index of refraction, which causes much internal reflection to occur in a diamond. This gives diamond the sparkly characteristic, making it valuable as a gem stone.

1. When a light wave enters a new medium and is refracted, there must be a change in the light wave's

   (1) color   (3) period
   (2) frequency   (4) speed   1 _____

2. The diagram below shows a ray of light passing from air into glass at an angle of incidence of 0°.

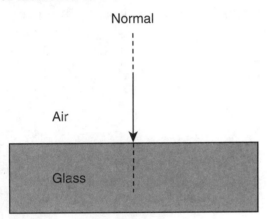

   Which statement best describes the speed and direction of the light ray as it passes into the glass?

   (1) Only speed changes.
   (2) Only direction changes.
   (3) Both speed and direction change.
   (4) Neither speed nor direction changes.

                                        2 _____

3. A beam of monochromatic light travels through flint glass, crown glass, Lucite, and water. The speed of the light beam is slowest in

   (1) flint glass   (3) Lucite
   (2) crown glass   (4) water   3 _____

Note: Question 4 has only three choices.

4. As yellow light ($f = 5.09 \times 10^{14}$ Hz) travels from zircon into diamond, the speed of the light

   (1) decreases
   (2) increases
   (3) remains the same   4 _____

5. The diagram below represents a ray of monochromatic light ($f = 5.09 \times 10^{14}$ Hz) passing from medium X ($n = 1.46$) into fused quartz.

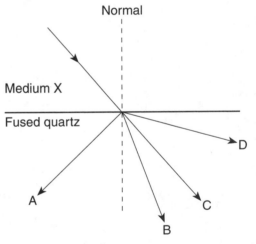

   Which path will the ray follow in the quartz?

   (1) A (2) B (3) C (4) D 5 _____

6. Which diagram best represents the path taken by a ray of monochromatic light as it passes from air through the materials shown?

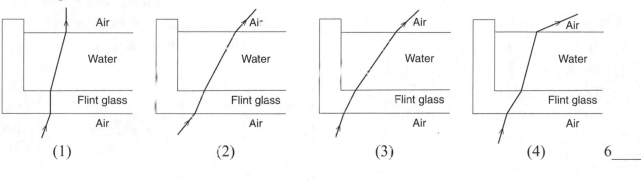

(1) (2) (3) (4) 6____

7. A ray of monochromatic light with a frequency of 5.09×10^{14} hertz is transmitted through four different media, listed to the right.

A. corn oil
B. ethyl alcohol
C. flint glass
D. water

Rank the four media from the one through which the light travels at the slowest speed to the one through which the light travels at the fastest speed. (Use the letters in front of each medium to indicate your answer.)

_____   _____   _____   _____
slowest speed   fastest speed

8. What occurs as a ray of light passes from air into water?

   (1) The ray must decrease in speed.
   (2) The ray must increase in speed.
   (3) The ray must decrease in frequency.
   (4) The ray must increase in frequency.

   8 _____

9. The diagram below shows a ray of light passing through two media.

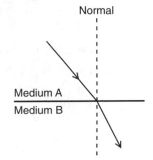

   When the wave travels from medium *A* into medium *B*, its speed

   (1) decreases
   (2) increases
   (3) remains the same   9 _____

10. The diagram below shows a ray of monochromatic light incident on an alcohol-flint glass interface.

   What occurs as the light travels from alcohol into flint glass?
   (1) The speed of the light decreases and the ray bends toward the normal.
   (2) The speed of the light decreases and the ray bends away from the normal.
   (3) The speed of the light increases and the ray bends toward the normal.
   (4) The speed of the light increases and the ray bends away from the normal.

   10 _____

11. Use the diagram below to arrive at your answer for the question below.

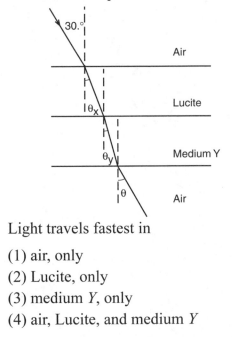

   Light travels fastest in

   (1) air, only
   (2) Lucite, only
   (3) medium *Y*, only
   (4) air, Lucite, and medium *Y*   11 _____

12. Which diagram best represents the behavior of a ray of monochromatic light in air incident on a block of crown glass?

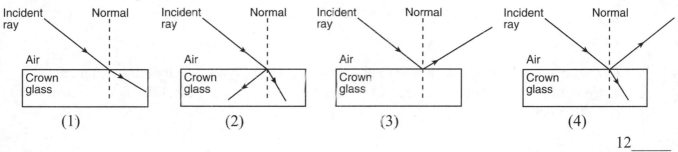

(1)   (2)   (3)   (4)

12_____

Base your answers to question 13 on the information and diagram below.

A ray of light ($f = 5.09 \times 10^{14}$ Hz) is incident on the boundary between air and an unknown material X at an angle of incidence of 55°, as shown. The absolute index of refraction of material X is 1.66.

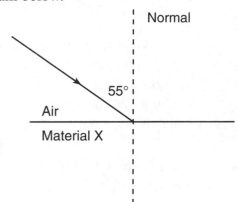

13. Identify a substance of which material X may be composed. [1]

_____

14. A ray of light traveling in air is incident on an airwater boundary as shown below. On the diagram below, draw the path of the ray in the water. [1]

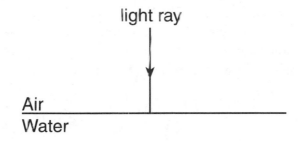

Charts

Page 181

Absolute Indices of Refraction
Answers
Set 1

1. **4** Refraction is defined as the change in direction of travel of a wave as it obliquely enters a medium in which its speed changes. As a wave enters a new medium, the color, frequency and period remain the same.

2. **1** As a wave travels from one medium into another with a different index of refraction, the speed will always change. For refraction (a change in the direction of travel of the wave) to occur, it must enter the new medium obliquely (strike the surface at an angle of incidence other then 0°). Since the wave enters along the normal (angle of incidence = 0°), no change in direction occurs.

3. **1** Under Waves, find the equation $n = c/v$. Solving for v, $v = c/n$. This shows that the speed of light in a material medium varies inversely with the absolute index of refraction of the medium. Therefore, the larger the absolute index of refraction, the slower the speed of light in that medium. Referring to the table of Absolute Indices of Refraction, flint glass has the largest value of the choices given.

4. **1** Under Waves, find the equation $n = c/v$. Solving for v gives $v = c/n$. This tells us that as the absolute index of refraction increases, the speed of light decreases. The speed of light in a vacuum (c) is a constant. Now find the absolute index of refraction of the substances on the Absolute Indices of Refraction table (zircon: $n = 1.92$, diamond: $n = 2.42$). As light travels from zircon onto diamond, the absolute index of refraction of the medium increases and the speed of light decreases.

5. **3** From the table of Absolute Indices of Refraction, the index of refraction of fused quartz is 1.46. This is the same as the index of medium X. The ray does not undergo a change in speed as it enters the fused quartz and therefore does not undergo any refraction or bending.

6. **2** Find the table of Absolute Indices of Refraction. The indices of refraction of the three substances are: air = 1.00, flint glass = 1.66, water = 1.33. As a ray of light enters a medium with a higher index of refraction, its speed decreases and it is refracted toward the normal. As a ray of light enters a medium with a smaller index of refraction, its speed increases and it is refracted away from the normal. Only choice 2 shows the proper changes in direction as the ray moves from one medium to the other.

7. Answer: $C\ A\ B\ D$

   Explanation: Under Waves, find the equation $n = c/v$. Solving for v gives $v = c/n$. Since c is a constant, the greater the index of refraction of a medium, the slower the speed of light in that medium. From the chart of Indices of Refraction in the reference table, the values for the four media given are: $A - 1.47$, $B - 1.36$, $C - 1.66$, $D - 1.33$. From these values, light travels slowest in C, then increases as you go from A to B to D.

Charts — Energy Level Diagrams

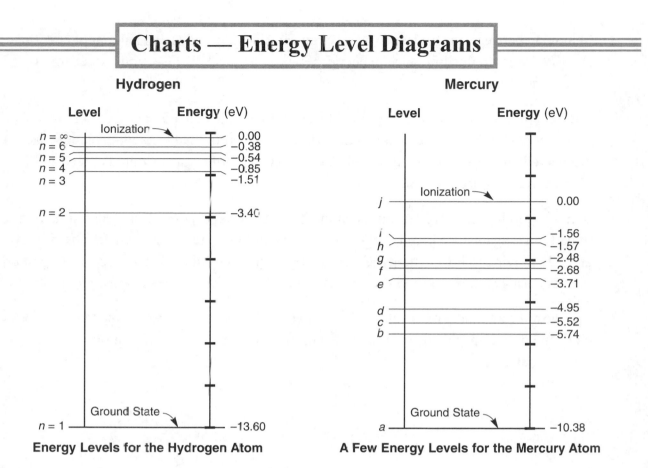

Hydrogen

Energy Levels for the Hydrogen Atom

Level	Energy (eV)
$n = \infty$ Ionization	0.00
$n = 6$	−0.38
$n = 5$	−0.54
$n = 4$	−0.85
$n = 3$	−1.51
$n = 2$	−3.40
$n = 1$ Ground State	−13.60

Mercury

A Few Energy Levels for the Mercury Atom

Level	Energy (eV)
j Ionization	0.00
i	−1.56
h	−1.57
g	−2.48
f	−2.68
e	−3.71
d	−4.95
c	−5.52
b	−5.74
a Ground State	−10.38

Overview:

In one modern theory of the structure of the atom (the Bohr model), electrons are considered to revolve around the nucleus in one of several concentric circular orbits, each corresponding to a particular energy for an electron. Each of these orbits is referred to as an energy level for that electron. As long as an electron remains in one of these orbits, its energy remains constant. However, an electron may absorb energy and move to a higher energy level or, if in a higher energy level, may lose or radiate energy and move to a lower energy level.

The Table:

The table shows the designation and energy values for the hydrogen atom and the mercury atom.
The ground state is the lowest energy level of an electron in an atom. Any energy level higher than the ground state is an excited state. The process of moving to a higher energy level is called excitation.
The ionization level is achieved when the electron is removed from the atom.
The energy values represent the potential energy of the electron with respect to the nucleus of the atom.

Additional Information:

- The potential energy of an electron an infinite distance from the nucleus is 0.00 eV. Since the energy of the electron decreases as it approaches the nucleus, the energy values of the energy levels is negative.

- The minimum amount of energy needed to remove an electron from an atom (move it to the 0.00 eV energy level) is called the ionization energy.

- In moving between energy levels, an electron can absorb or radiate an amount of energy precisely equal to the difference in energy between the levels. This amount of energy is called a quantum.

- The pattern of light absorbed or emitted by an atom is called an atomic spectrum. It is characteristic of a given element and may be used to identify that element. Helium was discovered on the Sun before it was discovered on the Earth by analyzing its spectrum in the light received from the Sun.

- In moving to higher energy levels, an atom forms an absorption spectrum, consisting of a series of dark lines against a colored background. The dark lines are the frequencies absorbed by the atom. In moving to lower energy levels, an emission spectrum is produced, consisting of bright lines against a dark background. The bright lines are the frequencies emitted by the atom.

- The most recent model of the atom, the cloud model or wave mechanical model, describes the electron not in a specific orbit around the nucleus but in a region of most probable location called a state.

1. In the currently accepted model of the atom, a fuzzy cloud around a hydrogen nucleus is used to represent the

    (1) electron's actual path, which is not a circular orbit
    (2) general region where the atom's proton is most probably located
    (3) general region where the atom's electron is most probably located
    (4) presence of water vapor in the atom

    1 _____

2. After electrons in hydrogen atoms are excited to the $n = 3$ energy state, how many different frequencies of radiation can be emitted as the electrons return to the ground state?

    (1) 1   (3) 3
    (2) 2   (4) 4   2 _____

3. A mercury atom in the ground state absorbs 16.78 electronvolts of energy and is ionized by losing an electron. How much kinetic energy does this electron have after the ionization?

    (1) 6.40 eV   (3) 10.38 eV
    (2) 9.62 eV   (4) 13.60 eV   3 _____

4. White light is passed through a cloud of cool hydrogen gas and then examined with a spectroscope. The dark lines observed on a bright background are caused by

    (1) the hydrogen emitting all frequencies in white light
    (2) the hydrogen absorbing certain frequencies of the white light
    (3) diffraction of the white light
    (4) constructive interference   4 _____

5. An electron in a mercury atom drops from energy level i to the ground state by emitting a single photon. This photon has an energy of

    (1) 1.56 eV   (3) 10.38 eV
    (2) 8.82 eV   (4) 11.94 eV   5 _____

6. A photon having an energy of 9.40 electronvolts strikes a hydrogen atom in the ground state. Why is the photon not absorbed by the hydrogen atom?

    (1) The atom's orbital electron is moving too fast.
    (2) The photon striking the atom is moving too fast.
    (3) The photon's energy is too small.
    (4) The photon is being repelled by electrostatic force.   6 _____

Base your answers to questions 7a, b and c on the information below.

An electron in a hydrogen atom drops from the $n = 3$ energy level to the $n = 2$ energy level.

7. a) What is the energy, in electronvolts, of the emitted photon? _____

   b) What is the energy, in joules, of the emitted photon? _____

   c) Calculate the frequency of the emitted radiation. [Show all work, including the equation and substitution with units.]

8. An electron in a mercury atom drops from energy level f to energy level c by emitting a photon having an energy of

    (1) 8.20 eV   (3) 2.84 eV
    (2) 5.52 eV   (4) 2.68 eV   8 _____

9. Moving electrons are found to exhibit properties of

    (1) particles, only
    (2) waves, only
    (3) both particles and waves
    (4) neither particles nor waves   9 _____

10. How much energy is required to move an electron in a mercury atom from the ground state to energy level h?

    (1) 1.57 eV   (3) 10.38 eV
    (2) 8.81 eV   (4) 11.95 eV   10 _____

11. Excited hydrogen atoms are all in the $n = 3$ state. How many different photon energies could possibly be emitted as these atoms return to the ground state?

    (1) 1   (3) 3
    (2) 2   (4) 4   11 _____

12. An electron in a hydrogen atom drops from the $n = 3$ energy level to the $n = 2$ energy level. The energy of the emitted photon is

    (1) 1.51 eV   (3) 3.40 eV
    (2) 1.89 eV   (4) 4.91 eV   12 _____

13. An electron in the c level of a mercury atom returns to the ground state. Which photon energy could *not* be emitted by the atom during this process?

    (1) 0.22 eV   (3) 4.86 eV
    (2) 4.64 eV   (4) 5.43 eV   13 _____

Base your answers to questions 14a and b on the information below.

In a mercury atom, as an electron moves from energy level i to energy level a, a single photon is emitted.

14. a) Determine the energy, in electronvolts, of this emitted photon. _____ eV

    b) Determine this photon's energy, in joules. _____ J

Copyright © 2011
Topical Review Book Company

Energy Level Diagrams
Answers
Set 1

1. **3** In the orbital or cloud model of the atom, only the most probable location of an electron around a nucleus can be given.

2. **3** The electron may drop from the $n = 3$ energy level to the $n = 1$ energy level in one step, emitting a single photon. It may also drop from the $n = 3$ energy level to the $n = 2$ energy level, emitting a single photon and then from the $n = 2$ energy level to the $n = 1$ energy level, emitting another single photon. These three possible photons have different energies and therefore different frequencies.

3. **1** Find the Energy Level Diagrams. Referring to that of Mercury, the ionization energy is 10.38 eV, the difference between the ground state (Level a) and the ionization state (Level j). The excess energy carried by the photon will be carried away by the electron as kinetic energy (16.78 eV – 10.38 eV = 6.40 eV).

4. **2** As the light passes through the hydrogen gas, the atoms will absorb certain frequencies causing the electron in the hydrogen atoms to move to higher energy levels. These frequencies or colors are therefore lacking in the observed spectrum and cause the black lines. This type of spectrum is called an absorption or dark line spectrum.

5. **2** Under Modern Physics, find the equation $E_{photon} = E_i - E_f$. Referring to the Energy Level Diagram for the Mercury atom, $E_{photon} = -1.56 \text{ eV} - (-10.38 \text{ eV}) = 8.82 \text{ eV}$.

6. **3** For a hydrogen atom in the ground state, the minimum energy a photon must possess to produce an electron transition would be the energy needed to excite the electron to the $n = 2$ energy level. Under Modern Physics, find the equation $E_{photon} = E_i - E_f$. Referring to the Energy Level Diagram for the Hydrogen atom, $E_{photon} = -13.6 \text{ eV} - (-3.40 \text{ eV}) = -10.2 \text{ eV}$. The negative sign indicates that the photon is being absorbed. Since the photon striking the Hydrogen atom has an energy equal to 9.40 eV, it is to small to produce excitation.

7. *a)* 1.89 eV

Explanation: Under Modern Physics, find the equation $E_{photon} = E_i - E_f$. Find the Energy Level Diagram for the Hydrogen atom. E_i is the energy of the $n = 3$ energy level and E_f is the energy of the $n = 2$ energy level. Substitution into the equation gives $E_{photon} = (-1.15 \text{ eV}) - (-3.40 \text{ eV}) = 1.89 \text{ eV}$. Be careful of the negative signs!

b) 3.02×10^{-19} J

Explanation: From the List of Physical Constants, $1 \text{ eV} = 1.60 \times 10^{-19}$ J. Use this to convert 1.89 eV (answer from question 7*a*) to joules using dimensional analysis.

$$1.89 \text{ eV} \times \frac{(1.60 \times 10^{-19} \text{ J})}{(1 \text{ eV})} = 3.02 \times 10^{-19} \text{ J}$$

c) $E_{photon} = hf \Rightarrow f = \dfrac{E_{photon}}{h} \Rightarrow f = \dfrac{3.02 \times 10^{-19} \text{ J}}{6.63 \times 10^{-34} \text{ J} \bullet \text{s}} \Rightarrow f = 4.56 \times 10^{14} \text{ Hz}$

Explanation: Under Modern Physics, find the equation $E_{photon} = hf$. h is Planck's constant and is found on the List of Physical Constants. Using the photon energy from question 7*b*, substitute into the equation and solve for f.

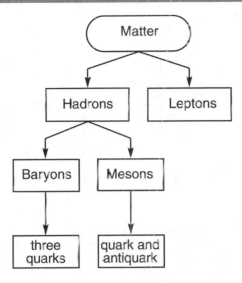

Overview:

Matter is defined as anything that has mass and takes up space. With the construction of more energetic particle accelerators, more fundamental building blocks of matter, referred to as elementary particles, have been discovered. As a result, a system of classifying these particles had to be developed.

The Table:

The building blocks of matter are divided into two groups, hadrons (from the Greek hadros, meaning strong) and leptons (from the Greek leptos, meaning small or light).

Hadrons interact by the strong nuclear force. Hadrons are in turn divided into baryons (from the Greek, barys, meaning heavy) and mesons (from the Greek mesos, meaning middle). Baryons include the particles known as the nucleons, which are the protons and neutrons. Baryons are in turn composed of three quarks (see Particles of the Standard Model). Mesons, which include pions and kaons, are composed of a quark and an antiquark.

Leptons include the familiar electron. There is no evidence that the electron has any internal structure similar to the hadrons.

Additional Information:

- According to a theory called quark confinement, a free quark cannot be observed. The force holding quarks together in protons and neutrons increases in strength rapidly as they separate from each other. This prevents their separation from one another.

- Mesons have masses between that of the electron and proton, hence their name.

Questions dealing with **Classification of Matter** can be found in the **Particles of the Standard Model** section.

Particles of the Standard Model

Quarks

Name	up	charm	top

Name — up — charm — top

Symbol — u — c — t

Charge — $+\frac{2}{3}$ e — $+\frac{2}{3}$ e — $+\frac{2}{3}$ e

down	strange	bottom
d	s	b
$-\frac{1}{3}$ e	$-\frac{1}{3}$ e	$-\frac{1}{3}$ e

Leptons

electron	muon	tau
e	μ	τ
$-1e$	$-1e$	$-1e$

electron neutrino	muon neutrino	tau neutrino
ν_e	ν_μ	ν_τ
0	0	0

Note: For each particle there is a corresponding antiparticle with a charge opposite that of its associated particles.

Overview:

The Standard Model is a theory that explains the composition of matter and how its component particles interact with each other. It is highly mathematical and is based on the interpretation of experimental observations and results since the particles are too small to be seen. There are over 100 different types of particles proposed in this model and they are grouped into three families, which are the quarks, leptons and force carriers. The table shown deals with the quarks and leptons.

The Table:

The Name, Symbol and Charge are given for the quarks and leptons.

The names given to the quarks are arbitrary and have no literal meaning. They simply identify certain properties of the quarks. Notice that the quarks possess fractional charges. They may be combined with each other to form hadrons.

The most familiar lepton is the ordinary electron. Leptons also include the neutrinos, which are electrically neutral, massless particles present in cosmic rays and are emitted by some radioactive elements. Having no charge or mass, they are extremely difficult to detect and pass through matter as if it was not present. There is a neutrino corresponding to each lepton.

This table is used with Classification of Matter to answer question dealing with the particles that make up matter.

Additional Information:

- Due to quark confinement (see Classification of Matter - Additional information), the fact that quarks possess fractional elementary charges does not violate the fact that the elementary charge is the minimum charge found on an isolated particle in nature.

- For each particle, there is an associated antiparticle having the same mass but an opposite charge. Some particles are their own antiparticles, as in the case of the neutron, which is neutral. Antiparticles are composed of antiquarks.

- A bar over a symbol indicates an antiparticle.

Charts

1. What fundamental force holds quarks together to form particles such as protons and neutrons?

   (1) electromagnetic force
   (2) gravitational force
   (3) strong force
   (4) weak force   1 _____

2. Baryons may have charges of

   (1) $+1e$ and $+\frac{4}{3}e$

   (2) $+2e$ and $+3e$

   (3) $-1e$ and $+1e$

   (4) $-2e$ and $-\frac{2}{3}e$ 2 _____

3. A lithium atom consists of 3 protons, 4 neutrons, and 3 electrons. This atom contains a total of

   (1) 9 quarks and 7 leptons
   (2) 12 quarks and 6 leptons
   (3) 14 quarks and 3 leptons
   (4) 21 quarks and 3 leptons   3 _____

4. A top quark has an approximate charge of

   (1) -1.07×10^{-19} C
   (3) $+1.07 \times 10^{-19}$ C
   (2) -2.40×10^{-19} C
   (4) $+2.40 \times 10^{-19}$ C   4 _____

5. Which combination of quarks could produce a neutral baryon?

   (1) cdt   (3) cdb
   (2) cts   (4) cdu   5 _____

Base your answers to questions 6a and b on the information below.

   A lambda particle consists of an up, a down, and a strange quark.

6. a) A lambda particle can be classified as a

   (1) baryon   (3) meson
   (2) lepton   (4) photon   6 a) _____

   b) What is the charge of a lambda particle in elementary charges?

   _____

Base your answers to questions 7a and b on the table below, which shows data about various subatomic particles.

Subatomic Particle Table

Symbol	Name	Quark Content	Electric Charge	Mass (GeV/c²)
p	proton	uud	+1	0.938
$\bar{p}$	antiproton	$\bar{u}\bar{u}\bar{d}$	−1	0.938
n	neutron	udd	0	0.940
λ	lambda	uds	0	1.116
Ω⁻	omega	sss	−1	1.672

7. a) Which particle listed on the table has the opposite charge of, and is more massive than, a proton?

   _____

   b) All the particles listed on the table are classified as

   _____

8. A meson may not have a charge of

    (1) +1e   (3) 0e
    (2) +2e   (4) −1e   8 _____

9. Protons and neutrons are examples of

    (1) positrons   (3) mesons
    (2) baryons   (4) quarks   9 _____

10. The charge of an antistrange quark is approximately

    (1) $+5.33 \times 10^{-20}$ C
    (2) -5.33×10^{-20} C
    (3) $+5.33 \times 10^{20}$ C
    (4) -5.33×10^{20} C   10 _____

11. According to the Standard Model of Particle Physics, a meson is composed of

    (1) a quark and a muon neutrino
    (2) a quark and an antiquark
    (3) three quarks
    (4) a lepton and an antilepton   11 _____

12. A particle unaffected by an electric field could have a quark composition of

    (1) *css*   (3) *udc*
    (2) *bbb*   (4) *uud*   12 _____

13. Which combination of quarks would produce a neutral baryon?

    (1) uud   (3) $\overline{u}\overline{u}d$
    (2) udd   (4) $\overline{u}dd$ 13 _____

14. The particles in a nucleus are held together primarily by the

    (1) strong force
    (2) gravitational force
    (3) electrostatic force
    (4) magnetic force   14 _____

15. The tau neutrino, the muon neutrino, and the electron neutrino are all

    (1) leptons   (3) baryons
    (2) hadrons   (4) mesons   15 _____

16. A tau lepton decays into an electron, an electron antineutrino, and a tau neutrino, as represented in the reaction below.

$$\tau \rightarrow e + \overline{v}_e + v_\tau$$

On the equation below, show how this reaction obeys the Law of Conservation of Charge by indicating the amount of charge on each particle.

_____e → _____ e + _____ e + _____ e

Particles of the Standard Model
Answers – Set 1

1. **3** The force that holds quarks together to form protons and neutrons (baryons) is the strong force.

2. **3** Locate the Classification of Matter. Baryons are a class of hadrons. Baryons include protons and neutrons (nucleons). There are no fractional charges observed for hadrons, which eliminates answers 1 and 4. The chart also indicates that baryons are composed of three quarks. Now go to the Particles of the Standard Model chart. Using any combination of 3 quarks and their corresponding charges, total charges of +2e, –1e and +1e are possible.
   Examples: $(+2/3e) + (+2/3e) + (+2/3e) = (+6/3e) = +2e$ and
   $(+2/3e) + (+2/3e) + (–1/3e) = (+4/3e) + (–1/3e) = (+3/3e) = +1e.$
   A charge of +3e is not possible.

3. **4** Refer to the Classification of Matter and Particles of the Standard Model tables. Protons and neutrons are baryons. Each baryon is composed of three quarks. Therefore, the number of quarks is $(3 \times 3) = 9$ for the protons and $(3 \times 4) = 12$ for the neutrons. Electrons are leptons. The number of electrons in a neutral atom equals the number of protons. Therefore, a lithium atom contains a total of $(9 + 12) = 21$ quarks and 3 leptons.

4. **3** On the Particles of the Standard Model table, the charge of the top quark is +2/3 e. On the List of Physical Constants, the elementary charge (e) is 1.60×10^{-19} C (1 e $= 1.60 \times 10^{-19}$ C). Use
   this to convert elementary charge units to coulombs $\frac{+2e}{3} \times \frac{(1.60 \times 10^{-19}\ \text{C})}{(1\ \text{e})} = +1.07 \times 10^{-19}\ \text{C}$.

5. **3** In the section Classification of Matter, it is found that baryons are composed of three quarks. Under Particles of the Standard Model, the names and charges of quarks are given. Only the combination of the c, d and b quarks would produce a neutral baryon: $\left(+\frac{2}{3}\right) + \left(-\frac{1}{3}\right) + \left(-\frac{1}{3}\right) = 0$.

6. *a)* 1 Explanation: Find the Classification of Matter chart. On the bottom left are particles composed of three quarks. These particles are classified as baryons.

   b) 0e *or* neutral
   Explanation: Find the Particles of the Standard Model chart. The names and charges of the quarks are given on the chart. The charge of the lambda particle is the sum of the charges of the three quarks.
   charge = (charge of up quark) + (charge of down quark) + (charge of strange quark)
   charge = $(+2/3$ e$) + (–1/3$ e$) + (–1/3$ e$) = 0$e

7. *a)* omega
   Explanation: The charge and mass of a proton, as given on the table, are +1 and 0.938 GeV/c², respectively. The particle in question must have a charge of –1 and a mass greater than 0.938 GeV/c². This is the omega particle.

   b) baryons
   Explanation: The table indicates that all of the particles listed on the table are composed of quarks. Find the Classification of Matter chart. The hadrons are composed of quarks.

Charts – Circuit Symbols

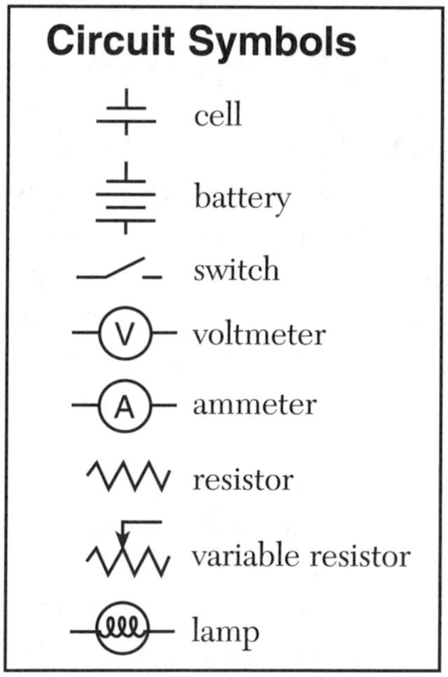

Circuit Symbols

⊥	cell
⊥	battery
⟋	switch
—(V)—	voltmeter
—(A)—	ammeter
⋀⋁⋀	resistor
⋀⋁⋀	variable resistor
—(lll)—	lamp

Overview:

There are two basic ways to connect components in an electrical circuit, series and parallel.
In a series circuit, the components are connected such that there is a single conducting path.
In a parallel circuit, two or more components are connected such that they provide separate conducting paths.

The Table:

The easiest way to show the way components are connected in an electric circuit is by a diagram or schematic. This table shows the accepted symbols used in circuit diagrams.

Additional Information:

- A battery is simply a combination of cells.

- A voltmeter is connected in parallel with a component to measure the potential difference or voltage across it.

- An ammeter is connected in series with a component to measure the current through it.

- A switch is placed in series with the device it controls.

- Any component of an electric circuit offers resistance to current in a circuit.

1. What must be inserted between points *A* and *B* to establish a steady electric current in the incomplete circuit represented in the diagram below?

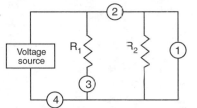

   (1) switch
   (2) voltmeter
   (3) magnetic field source
   (4) source of potential difference   1 _____

2. Two resistors are connected to a source of voltage as shown in the diagram below.

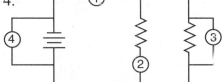

   At which position should an ammeter be placed to measure the current passing only through resistor R_1?

   (1) 1   (2) 2   (3) 3   (4) 4   2 _____

3. In the electric circuit diagram below, possible locations of an ammeter and a voltmeter are indicated by circles 1, 2, 3, and 4.

   Where should an ammeter be located to correctly measure the total current and where should a voltmeter be located to correctly measure the total voltage?

   (1) ammeter at 1 and voltmeter at 4
   (2) ammeter at 2 and voltmeter at 3
   (3) ammeter at 3 and voltmeter at 4
   (4) ammeter at 1 and voltmeter at 2   3 _____

4. Which circuit diagram shows voltmeter V and ammeter A correctly positioned to measure the total potential difference of the circuit and the current through each resistor?

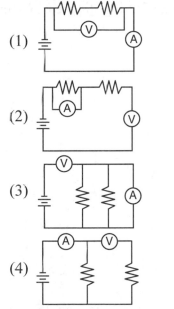

   (1)

   (2)

   (3)

   (4)

        4 _____

5. A 6.0-ohm lamp requires 0.25 ampere of current to operate. In which circuit below would the lamp operate correctly when switch S is closed?

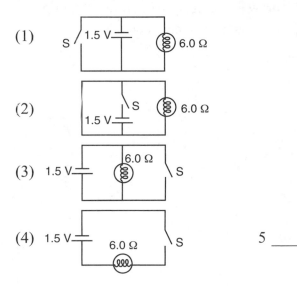

   (1) S 1.5 V 6.0 Ω

   (2) S 1.5 V 6.0 Ω

   (3) 1.5 V 6.0 Ω S

   (4) 1.5 V 6.0 Ω S   5 _____

Base your answers to question 6 on the information below.

An electric circuit contains two 3.0-ohm resistors connected in parallel with a battery. The circuit also contains a voltmeter that reads the potential difference across one of the resistors.

6. In the space provided below, draw a diagram of this circuit, using the symbols from the Reference Tables for Physical Setting/Physics. [Assume availability of any number of wires of negligible resistance.]

Base your answers to question 7 on the information and data table below.

Three lamps were connected in a circuit with a battery of constant potential. The current, potential difference, and resistance for each lamp are listed in the accompanying data table. [There is negligible resistance in the wires and the battery.]

7. Using the circuit symbols found in the Reference Tables for Physical Setting/Physics, draw a circuit showing how the lamps and battery are connected.

	Current (A)	Potential Difference (V)	Resistance (Ω)
lamp 1	0.45	40.1	89
lamp 2	0.11	40.1	365
lamp 3	0.28	40.1	143

8. In which circuit would current flow through resistor R_1, but not through resistor R_2 while switch S is open?

9. In which circuit represented below are meters properly connected to measure the current through resistor R_1 and the potential difference across resistor R_2?

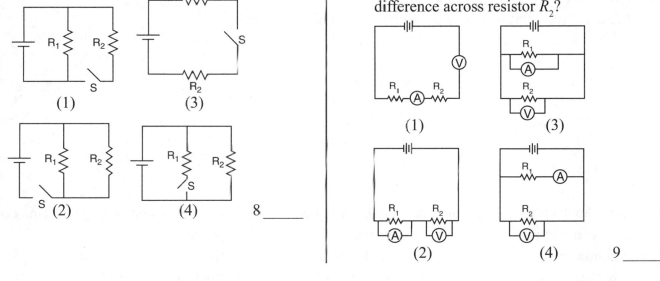

8 _____

9 _____

10. An electric circuit contains a source of potential difference and 5-ohm resistors that combine to give the circuit an equivalent resistance of 15 ohms. In the space below, draw a diagram of this circuit using circuit symbols given in the Reference Tables for Physical Setting/Physics. [Assume the availability of any number of 5-ohm resistors and wires of negligible resistance.]

Base your answers to question 11 on the information below.

A 5.0-ohm resistor, a 10.0-ohm resistor, and a 15.0-ohm resistor are connected in parallel with a battery. The current through the 5.0-ohm resistor is 2.4 amperes.

11. In the space below, using the circuit symbols found in the Reference Tables for Physical Setting/Physics, draw a diagram of this electric circuit.

12. Your school's physics laboratory has the following equipment available for conducting experiments:

accelerometers	lasers	stopwatches
ammeters	light bulbs	thermometers
bar magnets	meter sticks	voltmeters
batteries	power supplies	wires
electromagnets	spark timers	

Explain how you would find the resistance of an unknown resistor in the laboratory. Your explanation must include:

a) Measurements required

b) Equipment needed

c) Complete circuit diagram

d) Any equation(s) needed to calculate the resistance

1. 4   To produce a flow of charge or an electric current in a circuit, there must be an electrical potential difference between two points in that circuit.

2. 3   An ammeter must be placed in series with the device that it is to measure the current through. Position 3 is in series with resistor R_1.

3. 1   To measure total current, the ammeter must be in series with the parallel combination of resistors (position 1) and the voltmeter must be in parallel with the battery (position 4).

4. 1   A voltmeter must be placed in parallel with the device it measures the potential difference across and an ammeter must be placed in series with the device that it measures the current through. In diagram 1, the voltmeter is placed in parallel across both resistors, which are in series with each other, and therefore measures the total potential difference applied to the circuit. In diagram 1, the ammeter is in series with both resistors and therefore measures the current through each resistor.

5. 4   To operate properly, the lamp, switch and current source (the battery) must be connected in series with each other. Only circuit 4 shows this type of connection.

6.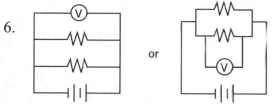

or

Explanation- To be connected in parallel with the battery, each resistor must provide a separate path for current. To measure the potential difference across one of the resistors, the voltmeter must be connected in parallel with that resistor.

7.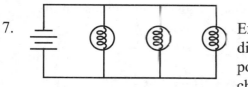

Explanation: Note on the table that the current through each lamp is different, indicating that they cannot be connected in series. The potential difference across each lamp is the same, which is characteristic of a parallel circuit. The lamps must be drawn connected in parallel with the voltage source.

Resistivities at 20°C

Material	Resistivity ($\Omega \cdot m$)
Aluminum	2.82×10^{-8}
Copper	1.72×10^{-8}
Gold	2.44×10^{-8}
Nichrome	$150. \times 10^{-8}$
Silver	1.59×10^{-8}
Tungsten	5.60×10^{-8}

Overview:

Resistivity (ρ) is a measure of the amount of opposition a material offers to a flow of charge through it. A flow of charge through a material is called an electric current and the opposition it offers to the current is called resistance. It is a basic property of a material that can be used to help identify a specific material.

The Table:

The table gives the material and its resistivity in ohm-meters ($\Omega \cdot m$) at 20° C. The temperature is specified since resistivity is temperature dependent.

Additional Information:

- The resistivity of a material is a property of that material and remains constant at constant temperature.

- As the temperature increases, the resistivities of metals increases, causing an increase in the resistance of metals as their temperature increases.

- Silver, having the lowest resistivity, is the best metallic conductor.

- The relatively low value of the resistivity of copper and its low cost relative to silver makes it the material used most commonly for household wiring.

- The high value of the resistivity of nichrome causes it to have a high resistance to an electric current, changing most of the electrical energy into heat energy. Therefore, it is used as the heating element in toasters and electric heaters.

1. Aluminum, copper, gold, and nichrome wires of equal lengths of 1.0×10^{-1} meter and equal cross-sectional areas of 2.5×10^{-6} meter^2 are at 20.°C. Which wire has the greatest electrical resistance?

   (1) aluminum   (3) gold
   (2) copper   (4) nichrome   1 _____

Note: Question 3 has only three choices.

3. The diagram represents a lamp, a 10-volt battery, and a length of nichrome wire connected in series.

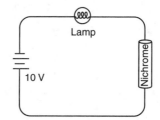

   As the temperature of the nichrome is decreased, the brightness of the lamp will

   (1) decrease
   (2) increase
   (3) remain the same   3 _____

2. Pieces of aluminum, copper, gold, and silver wire each have the same length and the same cross-sectional area. Which wire has the *lowest* resistance at 20°C?

   (1) aluminum   (3) gold
   (2) copper   (4) silver   2 _____

4. Which of the following materials would be best to use for a heating element?
   (1) aluminum   (3) nichrome
   (2) copper   (4) tungsten   4 _____

Note: Question 5 has only three choices.

5. A complete circuit is left on for several minutes, causing the connecting copper wire to become hot. As the temperature of the wire increases, the electrical resistance of the wire

   (1) decreases   (2) increases   (3) remains the same   5 _____

6. How does a superconductor differ from an ordinary conductor?

   _____

   _____

7. The accompanying graph shows the relationship between the potential difference across a metallic conductor and the electric current through the conductor at constant temperature T_1.

Which graph best represents the relationship between potential difference and current for the same conductor maintained at a higher constant temperature, T_2?

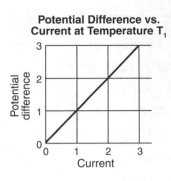

Potential Difference vs. Current at Temperature T₁

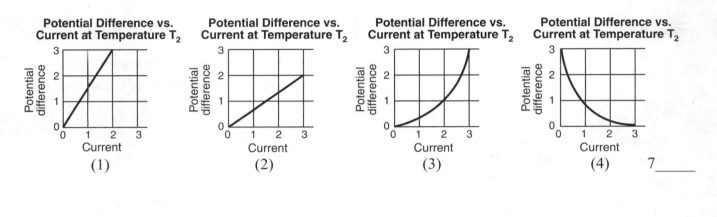

Potential Difference vs. Current at Temperature T₂

(1)　　　　(2)　　　　(3)　　　　(4)　　　7_____

═══════ **Resistivities at 20° C** ═══════

Answers

Set 1

1. 4　Find the equation $R = \rho L/A$. The larger the resistivity, the larger the resistance of an electrical conductor is. Referring to the table of Resistivities at 20° C, nichrome has the greatest resistivity.

2. 4　Under Electricity, find the equation $R = \rho L/A$. This equation shows that the resistance of a wire varies directly with the resistivity of the material. Find the Resistivities at 20° C in the reference table. Note the value for each material. For the same length and cross-sectional area, the wire with the smallest resistivity will have the smallest resistance. This is silver

3. 2　As the temperature of the nichrome wire decreases, its resistance decreases. At a constant potential difference, a decrease in the resistance of the circuit will cause an increase in the current in the circuit. Since the nichrome wire and lamp are connected in series, the current through the lamp increases, increasing the brightness of the lamp.

4. 3　Under Electricity, find the equation $R = \rho L/A$. The greater the resistivity, the greater the resistance a material offers to current through it and the greater the amount of heat the current produces. Referring to the table of Resistivities at 20° C, nichrome has the greatest resistivity.